Busy Ant Maths

Stretch and Challenge 3

A problem-solving, cross-curricular programme for children working above end-of-year expectations

Peter Clarke

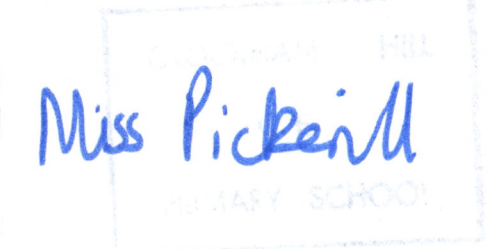

William Collins' dream of knowledge for all began with the publication of his first book in 1819.
A self-educated mill worker, he not only enriched millions of lives, but also founded a flourishing publishing house. Today, staying true to this spirit, Collins books are packed with inspiration, innovation and practical expertise. They place you at the centre of a world of possibility and give you exactly what you need to explore it.

Collins. Freedom to teach.

Published by Collins

An imprint of HarperCollinsPublishers
The News Building
1 London Bridge Street
London SE1 9GF

Browse the complete Collins catalogue at
www.collins.co.uk

© HarperCollins*Publishers* Limited 2016

10 9 8 7 6 5 4 3

ISBN 978-0-00-816732-5

Peter Clarke asserts his moral rights to be identified as the author of this work.

Any educational institution that has purchased one copy of this publication may make duplicate copies of pages 23–94 and 209–224 for use exclusively within that institution. Permission does not extend to reproduction, storage in a retrieval system or transmission in any form or by any means – electronic, mechanical, photocopying, recording or otherwise – of duplicate copies for lending renting or selling to any other user or institution without the prior consent, in writing, of the Publisher.

British Library Cataloguing in Publication Data
A Catalogue record for this publication is available from the British Library.

Commissioned by Fiona McGlade
Cover and series design by Kneath Associates
Template creation by Ken Vail Graphic Design
Illustrations by Mark Walker, Steve Evans, Gwyneth Williamson and QBS Learning
Typesetting by QBS Learning
Editing and proofreading by Alissa McWhinnie

Printed and bound by Printed by CPI Group (UK) Ltd, Croydon, CR0 4YY

Acknowledgements
Peter Clarke wishes to thank Brian Molyneaux for his valuable contribution to this publication.

Contents

Quick reference guide to *Stretch and Challenge 3*	4
Introduction	5
The features of *Stretch and Challenge*	7
A possible *Stretch and Challenge* teaching and learning sequence	10
Links to the Year 3 Mathematics National Curriculum Programme of Study and Attainment Targets	11
Cross-curricular links to the National Curriculum Programme of Study	19
Resources used in *Stretch and Challenge 3*	21
The Issues	23
Teacher's notes	95
Resource sheets	
Record of completion	209
My notes	211
Pupil self assessment booklet	213
Other Resource sheets	215

Quick reference guide to *Stretch and Challenge 3*

Domain(s)	Topic	Issue number	Teacher's notes page number
Number: – Number and place value	Number	1	96
	Number	2	99
	Number	3	103
	Number	4	106
Number: – Addition and subtraction	Addition	5	109
	Addition	6	112
	Subtraction	7	115
	Subtraction	8	118
Number: – Multiplication and division	Multiplication	9	121
	Multiplication	10	124
	Division	11	127
	Division	12	130
Number: – Addition and subtraction – Multiplication and division	Mixed operations	13	133
	Mixed operations	14	136
	Mixed operations	15	139
	Mixed operations	16	142
Number: – Fractions	Fractions	17	145
	Fractions	18	148
	Fractions	19	151
	Fractions	20	154
Measurement	Length	21	157
	Mass	22	160
	Capacity and volume	23	163
	Time	24	166
	Measurement	25	169
	Measurement	26	173
Geometry: – Properties of shapes	2-D shapes	27	177
	3-D shapes	28	180
	Symmetry	29	183
	Position and direction	30	187
	Movement and angle	31	190
	Geometry	32	193
	Geometry	33	196
Statistics	Statistics	34	199
	Statistics	35	202
	Statistics	36	205

Introduction

The National Curriculum emphasises the importance of all children mastering the programme of study taught each year and discourages the acceleration of children into content from subsequent years.

The National Curriculum states: *'The expectation is that the majority of pupils will move through the programmes of study at broadly the same pace. However, decisions about when to progress should always be based on the security of pupils' understanding and their readiness to progress to the next stage. Pupils who grasp concepts rapidly should be challenged through being offered rich and sophisticated problems before any acceleration through new content. Those who are not sufficiently fluent with earlier material should consolidate their understanding, including through additional practice, before moving on.'* [1]

However, the National Curriculum also goes on to say that: *'Within each key stage, schools [therefore] have the flexibility to introduce content earlier or later than set out in the programme of study. In addition, schools can introduce key stage content during an earlier key stage, if appropriate.'* [2]

Stretch and Challenge aims to provide support in meeting the needs of those children who are exceeding age-related expectations by providing a range of problem-solving and cross-curricular activities designed to enrich and deepen children's mathematical knowledge, skills and understanding.

The series provides opportunities for children to reason mathematically and to solve increasingly complex problems, doing so with fluency, as described in the aims of the National Curriculum:

'The National Curriculum for mathematics aims to ensure that all pupils:

- become **fluent** *in the fundamentals of mathematics, including through varied and frequent practice with increasingly complex problems over time, so that pupils develop conceptual understanding and the ability to recall and apply knowledge rapidly and accurately*

- **reason mathematically** *by following a line of enquiry, conjecturing relationships and generalisations, and developing an argument, justification or proof using mathematical language*

- *can* **solve problems** *by applying their mathematics to a variety of routine and non-routine problems with increasing sophistication, including breaking down problems into a series of simpler steps and persevering in seeking solutions.'* [3]

Stretch and Challenge has been designed to provide:

- a flexible 'dip-in' resource that can easily be adapted to meet the needs of individual children, and different classroom and school organisational arrangements

- enrichment activities that require children to use and apply their mathematical knowledge, skills and understanding to reason mathematically and to solve increasingly complex problems

- mathematical activities linked to the entire primary curriculum, thereby ensuring a range of cross-curricular contexts

- an easy-to-use bank of activities to save teachers time in thinking up new enrichment activities

- an interesting, unique and consistent approach to presenting enrichment activities to children.

The *Stretch and Challenge* series consists of six packs, also available digitally on Collins Connect, one for each year group from Year 1 to Year 6.

---

[1] Mathematics programmes of study: key stages 1 and 2 National Curriculum in England, September 2013, page 3
[2] Mathematics programmes of study: key stages 1 and 2 National Curriculum in England, September 2013, page 4
[3] Mathematics programmes of study: key stages 1 and 2 National Curriculum in England, September 2013, page 3

Printed resources

Containing:

- Pupil activity booklets (Issues)
- Teacher's notes
- Resource sheets

Online resources at connect.collins.co.uk

Containing editable:

- Pupil activity booklets (Issues)
- Teacher's notes
- Resource sheets

It is envisaged that the activities in *Stretch and Challenge* will be used by either individuals or pairs of children. However, given the flexible nature of the resource, if appropriate, children can work in groups. The activities are intended to be used:

- as additional work to be done once children have finished other set work
- by those children who grasp concepts rapidly and need to be challenged through rich and sophisticated problems
- as in-depth work that is to be undertaken over a prolonged period of time, such as during the course of several lessons, a week or a particular unit of work
- as a resource for promoting mathematical reasoning and problem solving and developing independent thinking and learning
- as a springboard for further investigations into mathematics based on the children's suggestions.

Introduction

The features of *Stretch and Challenge*

Pupil activity booklets (Issues)

- Each of the 36 Issues in *Stretch and Challenge 3* consists of a four-page A5 pupil activity booklet (to be printed double sided onto one sheet of A4 paper).

- The 36 Issues cover the different domains and attainment targets of the Mathematics National Curriculum Programme of Study (see pages 11–18).

- The Issues have been designed to resemble a newspaper, with each of the Issues consisting of between five and eight different activities, all related to the same mathematical topic.

- It is important to note that children are not expected to complete all the activities in an Issue nor work their way through an Issue from beginning to end. For many children not all of the activities offered in an Issue will be appropriate. When choosing which activities a child is to complete, teachers should ensure that the activities do not accelerate the child into mathematical content they may not be familiar with, or are unable to reason more deeply in order to develop a conceptual understanding. Rather, activities should be chosen on the basis that they engage the child in reasoning and the development of mathematical thinking, as well as enriching and deepening the child's mathematical knowledge, skills and understanding.

- The terms 'Issue' and 'Volume' have been used rather than 'Unit' and 'Year group' because they are in keeping with the newspaper theme.

Types of activities

- Each of the 36 Issues in *Stretch and Challenge 3* are designed to deepen children's mathematical knowledge, skills and understandings, and enhance their use and application of mathematics. There are four different types of 'using and applying' activities in the series:

    What's the Problem?   The Puzzler

    Looking for Patterns   Let's Investigate

- Alongside developing children's problem-solving skills, the series also provides activities with cross-curricular links to other subjects in the primary curriculum. The following shows the *Stretch and Challenge* features and its corresponding primary curriculum subject.

Curriculum subject	*Stretch and Challenge* feature
English	The Language of Maths
Science	Focus on Science
Computing	Technology Today
Geography	Around the World
History	In the Past
Mathematics	Famous Mathematicians
Art and design / Music	The Arts Roundup
Design and Technology	Construct
Physical Education	Sports Update

- As well as the features mentioned above, other regular features in *Stretch and Challenge* include:

    Money Matters   At Home (home–school link activities).

- A chart showing the link between the Issues, the *Stretch and Challenge* features and cross-curricular links can be found on pages 19 and 20.

- Inquisitive ant is a recurring feature of the series. In each Issue there is an ant holding a mathematical word or symbol. Children locate the ant and write about the meaning of the word or symbol.

Introduction

Teacher's notes

Each of the 36 Issues includes a set of teacher's notes, including answers.

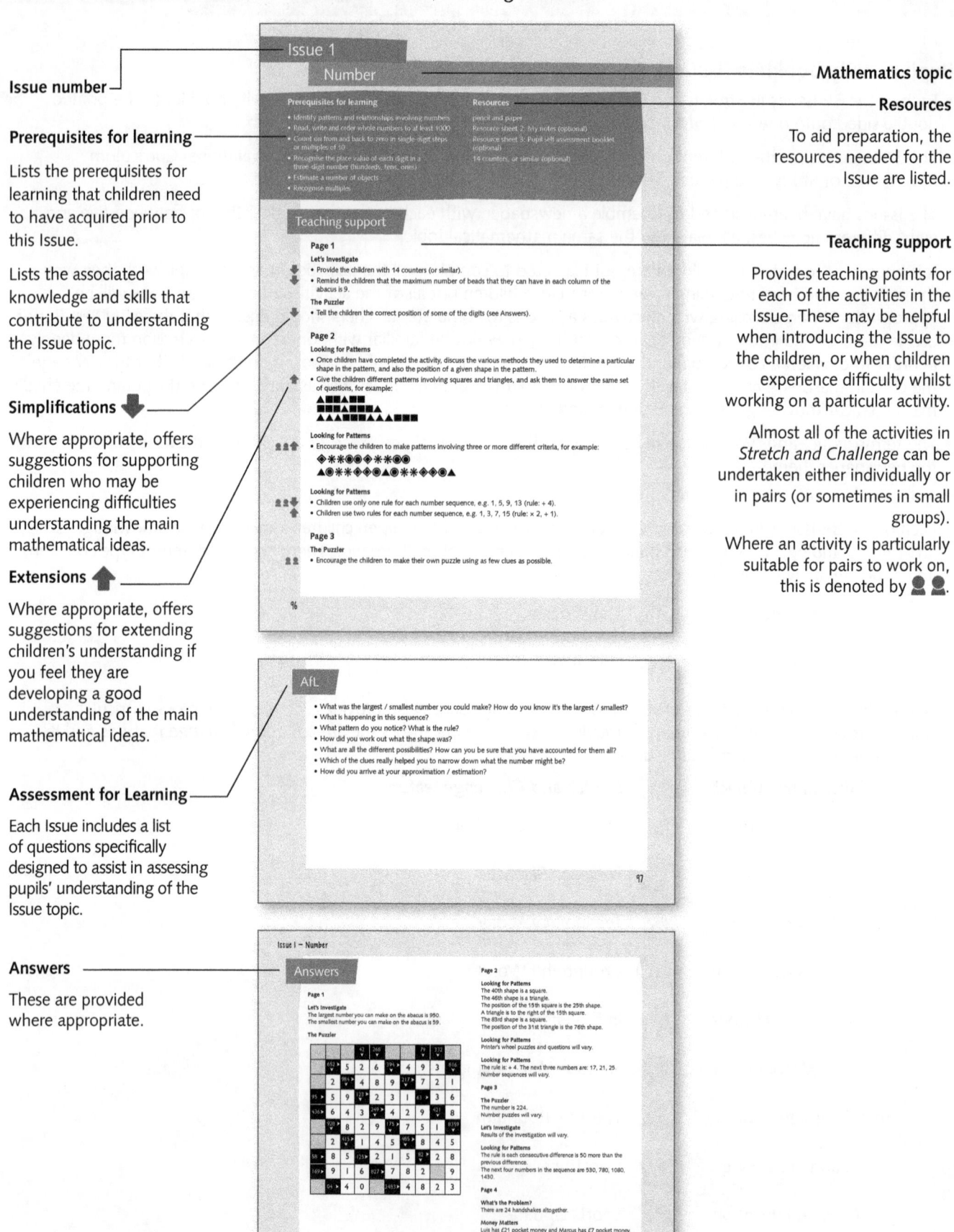

Issue number

Prerequisites for learning

Lists the prerequisites for learning that children need to have acquired prior to this Issue.

Lists the associated knowledge and skills that contribute to understanding the Issue topic.

Simplifications

Where appropriate, offers suggestions for supporting children who may be experiencing difficulties understanding the main mathematical ideas.

Extensions

Where appropriate, offers suggestions for extending children's understanding if you feel they are developing a good understanding of the main mathematical ideas.

Assessment for Learning

Each Issue includes a list of questions specifically designed to assist in assessing pupils' understanding of the Issue topic.

Answers

These are provided where appropriate.

Mathematics topic

Resources

To aid preparation, the resources needed for the Issue are listed.

Teaching support

Provides teaching points for each of the activities in the Issue. These may be helpful when introducing the Issue to the children, or when children experience difficulty whilst working on a particular activity.

Almost all of the activities in *Stretch and Challenge* can be undertaken either individually or in pairs (or sometimes in small groups).

Where an activity is particularly suitable for pairs to work on, this is denoted by 👥.

Introduction

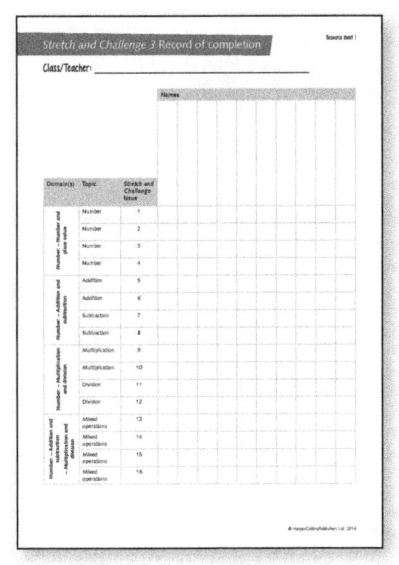

Record of completion

To assist in keeping a record of which Issues children have completed.

Once a child has completed an Issue you could either put a tick or write the date in the corresponding box.

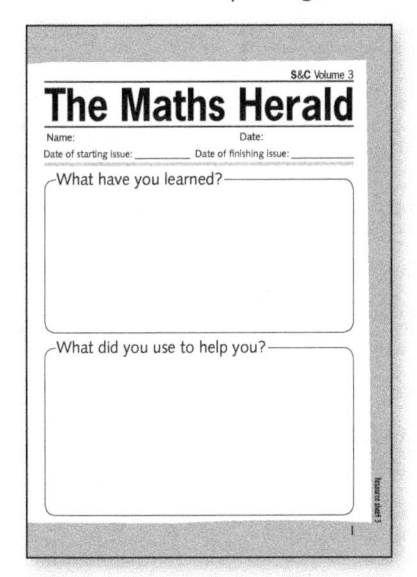

Pupil self assessment booklet

Each Resource Pack in the *Stretch and Challenge* series includes an age-appropriate pupil self assessment A5 booklet (to be printed double sided onto one sheet of A4 paper).

This booklet is a generic sheet that can be used for any, or all, of the 36 Issues in the Resource Pack.

The booklet is designed to provide children with an opportunity to undertake some form of self assessment once they have completed the Issue.

After the children have completed the booklet, discuss with them what they have written.

This can then be kept, together with the child's copy of the Issue and their working out and answers, including, if appropriate, 'My notes'.

My notes

The pupil activity booklets have been designed to resemble a newspaper. This means that quite often there is insufficient space in the booklets for children to show their working and answers.

You may decide to simply provide children with pencil and paper to record their work or an exercise book that they use as their '*Stretch and Challenge* Journal'. Alternatively, you could provide them with a copy of the A5 booklet: 'My notes' (to be printed doubled sided onto one sheet of A4 paper). This can then be kept, together with the child's copy of the Issue and, if appropriate, their 'Pupil self assessment booklet.'

Whichever method you choose for the children to record their working and answers, i.e. on sheets of paper, using a 'My notes' booklet, in an exercise book, or any other method, children need to be clear and systematic in their recording.

Other Resource sheets

For some of the activities, children are required to use a specific Resource sheet.

These are included both in the back of this Resource Pack and the online resources.

Introduction

A possible *Stretch and Challenge* teaching and learning sequence

As the diagram on the right illustrates, the process of learning about mathematics can be thought of as the interrelationship between knowledge, understanding and application.

A suggested teaching sequence for working with children based on this model and using the activities in *Stretch and Challenge* is given below.

A complete sequence may occur during a particular lesson if the activity given is designed to be completed during the course of the lesson. Alternatively, the teaching and learning sequence may extend for a longer period of time if the activities are to be completed over the course of several lessons, a week or during a particular unit of work.

Puzzle pieces labelled: Constructing meaning, Transferring knowledge, Using & applying, Understanding.

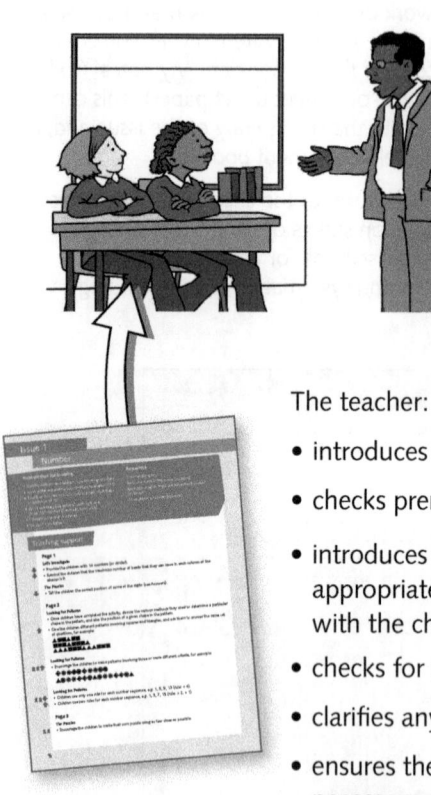

Briefing

The teacher:

- introduces the topic
- checks prerequisites for learning
- introduces the activity including, where appropriate, reading through the activity with the children
- checks for understanding
- clarifies any misconceptions
- ensures there is easy access to any necessary resources.

Working

- Children work individually, in pairs, or if appropriate, in groups.
- Teacher acts only as a 'guide-on-the-side'.

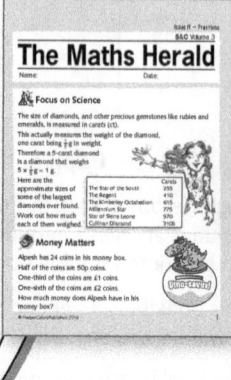

De-briefing

The child / children:

- reports back to others
- reflects on their learning
- identifies the 'next step'.

The teacher evaluates learning.

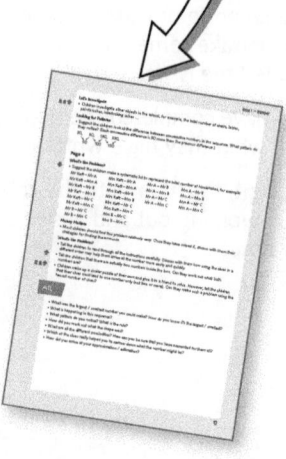

10

Links to the Year 3 Mathematics National Curriculum Programme of Study and Attainment Targets

Stretch and Challenge Issue

Number – Number and place value	1	2	3	4	5	6	7	8	9	10	11	12	13	14	15	16	17	18	19	20	21	22	23	24	25	26	27	28	29	30	31	32	33	34	35	36
• count from 0 in multiples of 4, 8, 50 and 100; find 10 or 100 more or less than a given number	•	•	•	•																																
• recognise the place value of each digit in a three-digit number (hundreds, tens, ones)	•	•	•	•		•		•																												
• compare and order numbers up to 1000	•	•	•																																	
• identify, represent and estimate numbers using different representations	•	•	•																																	
• read and write numbers up to 1000 in numerals and in words	•	•		•	•	•	•	•	•	•	•	•	•	•																						
• solve number problems and practical problems involving these ideas	•	•			•	•	•	•	•																											
Notes and guidance (non-statutory)																																				
Pupils now use multiples of 2, 3, 4, 5, 8, 10, 50 and 100.	•	•	•	•	•	•	•	•	•	•	•	•	•																							
They use larger numbers to at least 1000, applying partitioning related to place value using varied and increasingly complex problems, building on work in year 2 (for example, 146 = 100 + 40 and 6, 146 = 130 + 16).	•	•	•	•	•	•	•	•	•	•	•	•	•																							
Using a variety of representations, including those related to measure, pupils continue to count in ones, tens and hundreds, so that they become fluent in the order and place value of numbers to 1000.	•			•																																

Introduction

Number – Addition and subtraction

	Stretch and Challenge Issue
	1 2 3 4 5 6 7 8 9 10 11 12 13 14 15 16 17 18 19 20 21 22 23 24 25 26 27 28 29 30 31 32 33 34 35 36

- add and subtract numbers mentally, including:
  - a three-digit number and ones
  - a three-digit number and tens
  - a three-digit number and hundreds

  • columns 5, 6, 7, 8, 12, 13, 14, 15

- add and subtract numbers with up to three digits, using formal written methods of columnar addition and subtraction

  • columns 5, 6, 7, 8, 12, 13, 14, 15

- estimate the answer to a calculation and use inverse operations to check answers

  • columns 5, 6, 7, 8, 12, 13, 14, 15

- solve problems, including missing number problems, using number facts, place value, and more complex addition and subtraction

  • columns 5, 6, 7, 8, 12, 13, 14, 15

Notes and guidance (non-statutory)

Pupils practise solving varied addition and subtraction questions. For mental calculations with two-digit numbers, the answers could exceed 100.

• columns 5, 6, 7, 8, 12, 13, 14, 15

Pupils use their understanding of place value and partitioning, and practise using columnar addition and subtraction with increasingly large numbers up to three digits to become fluent.

• columns 5, 6, 7, 8, 12, 13, 14, 15

12

Introduction

Stretch and Challenge Issue

	1	2	3	4	5	6	7	8	9	10	11	12	13	14	15	16	17	18	19	20	21	22	23	24	25	26	27	28	29	30	31	32	33	34	35	36
Number – Multiplication and division																																				
• recall and use multiplication and division facts for the 3, 4 and 8 multiplication tables									•	•	•	•	•	•	•																					
• write and calculate mathematical statements for multiplication and division using the multiplication tables that they know, including for two-digit numbers times one-digit numbers, using mental and progressing to formal written methods									•	•	•	•	•	•	•																					
• solve problems, including missing number problems, involving multiplication and division, including positive integer scaling problems and correspondence problems in which n objects are connected to m objects									•	•	•	•	•	•	•																					

Introduction

Number – Multiplication and division Continued
Notes and guidance (non-statutory)

	Stretch and Challenge Issue
	1 2 3 4 5 6 7 8 9 10 11 12 13 14 15 16 17 18 19 20 21 22 23 24 25 26 27 28 29 30 31 32 33 34 35 36
Pupils continue to practise their mental recall of multiplication tables when they are calculating mathematical statements in order to improve fluency. Through doubling, they connect the 2, 4 and 8 multiplication tables.	• • • • • • •
Pupils develop efficient mental methods, for example, using commutativity and associativity (for example, 4 × 12 × 5 = 4 × 5 × 12 = 20 × 12 = 240) and multiplication and division facts (for example, using 3 × 2 = 6, 6 ÷ 3 = 2 and 2 = 6 ÷ 3) to derive related facts (for example, 30 × 2 = 60, 60 ÷ 3 = 20 and 20 = 60 ÷ 3).	• • • • • • •
Pupils develop reliable written methods for multiplication and division, starting with calculations of two-digit numbers by one-digit numbers and progressing to the formal written methods of short multiplication and division.	• • • • • • •
Pupils solve simple problems in contexts, deciding which of the four operations to use and why. These include measuring and scaling contexts, (for example, four times as high, eight times as long etc.) and correspondence problems in which m objects are connected to n objects (for example, 3 hats and 4 coats, how many different outfits?; 12 sweets shared equally between 4 children; 4 cakes shared equally between 8 children).	• • • • • • •

14

Stretch and Challenge Issue

Number – Fractions	1	2	3	4	5	6	7	8	9	10	11	12	13	14	15	16	17	18	19	20	21	22	23	24	25	26	27	28	29	30	31	32	33	34	35	36
• count up and down in tenths; recognise that tenths arise from dividing an object into 10 equal parts and in dividing one-digit numbers or quantities by 10																																				
• recognise, find and write fractions of a discrete set of objects: unit fractions and non-unit fractions with small denominators																•	•	•	•																	
• recognise and use fractions as numbers: unit fractions and non-unit fractions with small denominators																•	•	•	•																	
• recognise and show, using diagrams, equivalent fractions with small denominators																	•	•	•																	
• add and subtract fractions with the same denominator within one whole [for example, $\frac{5}{7} + \frac{1}{7} = \frac{6}{7}$]																•	•	•	•																	
• compare and order unit fractions, and fractions with the same denominators																•	•	•	•																	
• solve problems that involve all of the above																•	•	•	•																	
Notes and guidance (non-statutory)																																				
Pupils connect tenths to place value, decimal measures and to division by 10.																•	•	•																		
They begin to understand unit and non-unit fractions as numbers on the number line, and deduce relations between them, such as size and equivalence. They should go beyond the [0, 1] interval, including relating this to measure.																		•																		
Pupils understand the relation between unit fractions as operators (fractions of), and division by integers.																•	•	•																		
They continue to recognise fractions in the context of parts of a whole, numbers, measurements, a shape, and unit fractions as a division of a quantity.																•	•	•																		
Pupils practise adding and subtracting fractions with the same denominator through a variety of increasingly complex problems to improve fluency.																•	•	•																		

Introduction

Stretch and Challenge Issue

Measurement

Objective	Issues
• measure, compare, add and subtract: lengths (m/cm/mm); mass (kg/g); volume/capacity (l/ml)	
• measure the perimeter of simple 2-D shapes	
• add and subtract amounts of money to give change, using both £ and p in practical contexts	1, 6, 8, 9, 10, 12, 13, 14, 15, 16, 17, 21, 22, 23, 24, 25, 26
• tell and write the time from an analogue clock, including using Roman numerals from I to XII, and 12-hour and 24-hour clocks	21, 22, 23, 24, 25, 26
• estimate and read time with increasing accuracy to the nearest minute; record and compare time in terms of seconds, minutes and hours; use vocabulary such as o'clock, a.m./p.m., morning, afternoon, noon and midnight	21, 22, 23, 24, 25, 26
• know the number of seconds in a minute and the number of days in each month, year and leap year	22, 23, 24, 25, 26
• compare durations of events [for example to calculate the time taken by particular events or tasks]	22, 23, 24, 25, 26

Notes and guidance (non-statutory)

Pupils continue to measure using the appropriate tools and units, progressing to using a wider range of measures, including comparing and using mixed units (for example, 1 kg and 200g) and simple equivalents of mixed units (for example, 5m = 500cm). — 21, 22, 23, 24, 25, 26

The comparison of measures includes simple scaling by integers (for example, a given quantity or measure is twice as long or five times as high) and this connects to multiplication. — 21, 22, 23, 24, 25, 26

Pupils continue to become fluent in recognising the value of coins, by adding and subtracting amounts, including mixed units, and giving change using manageable amounts. They record £ and p separately. The decimal recording of money is introduced formally in year 4. — 1, 6, 22, 23, 24, 25

Pupils use both analogue and digital 12-hour clocks and record their times. In this way they become fluent in and prepared for using digital 24-hour clocks in year 4. — 24, 25, 26

16

Geometry – Properties of shapes

Stretch and Challenge Issue	1	2	3	4	5	6	7	8	9	10	11	12	13	14	15	16	17	18	19	20	21	22	23	24	25	26	27	28	29	30	31	32	33	34	35	36
• draw 2-D shapes and make 3-D shapes using modelling materials; recognise 3-D shapes in different orientations and describe them																										•	•	•					•			
• recognise angles as a property of shape or a description of a turn																											•		•			•				
• identify right angles, recognise that two right angles make a half-turn, three make three quarters of a turn and four a complete turn; identify whether angles are greater than or less than a right angle																											•	•	•	•	•	•				
• identify horizontal and vertical lines and pairs of perpendicular and parallel lines																										•							•			

Notes and guidance (non-statutory)

Pupils' knowledge of the properties of shapes is extended at this stage to symmetrical and non-symmetrical polygons and polyhedra. Pupils extend their use of the properties of shapes. They should be able to describe the properties of 2-D and 3-D shapes using accurate language, including lengths of lines and acute and obtuse for angles greater or lesser than a right angle.

Pupils connect decimals and rounding to drawing and measuring straight lines in centimetres, in a variety of contexts.

| | | | | | | | | | | | | | | | | | | | | | | | | | | • | | • | | | • | | | | |

Introduction

Stretch and Challenge Issue

	1	2	3	4	5	6	7	8	9	10	11	12	13	14	15	16	17	18	19	20	21	22	23	24	25	26	27	28	29	30	31	32	33	34	35	36
Statistics																																				
• interpret and present data using bar charts, pictograms and tables																																	•	•	•	•
• solve one-step and two-step questions [for example, 'How many more?' and 'How many fewer?'] using information presented in scaled bar charts and pictograms and tables																																	•	•	•	•
Notes and guidance (non-statutory)																																				
Pupils understand and use simple scales (for example, 2, 5, 10 units per cm) in pictograms and bar charts with increasing accuracy.																																	•	•		•
They continue to interpret data presented in many contexts.																																	•	•		•

Cross-curricular links to the National Curriculum Programme of Study

Cross-curricular link and *Stretch and Challenge* Feature

Category	Feature	1	2	3	4	5	6	7	8	9	10	11	12	13	14	15	16
Mathematics	What's the Problem?	•		•		•	•		•	•		•	•	•	•	•	•
Mathematics	Looking for Patterns	•	•	•		•	•		•	•	•	•	•		•	•	
Mathematics	The Puzzler	•	•	•	•	•	•	•	•	•	•	•	•	•	•	•	•
Mathematics	Let's Investigate	•	•	•	•	•	•	•	•	•	•	•	•			•	•
English	The Language of Maths		•	•	•				•				•				•
Science	Focus on Science				•							•			•		
Computing	Technology Today				•						•					•	•
Geography	Around the World			•					•								
History	In the Past / Famous Mathematicians				•			•			•						
Art & Design / Music	The Arts Roundup			•									•	•			•
Design & Technology	Construct			•													
Physical Education	Sports Update		•	•								•		•		•	
Links to money	Money Matters	•				•			•	•				•	•		•
Links with home	At Home													•	•		

Issue Topics

Domain(s)	Topic	Issue
Number: – Number and place value	Number	1
	Number	2
	Number	3
	Number	4
Number: – Addition and subtraction	Addition	5
	Addition	6
	Subtraction	7
	Subtraction	8
Number: – Multiplication and division	Multiplication	9
	Multiplication	10
	Division	11
	Division	12
Number: – Addition and subtraction	Mixed operations	13
	Mixed operations	14
Number: – Multiplication and division	Mixed operations	15
	Mixed operations	16

Introduction

Cross-curricular link and *Stretch and Challenge* Feature																					
	Links with home	At Home			•		•				•		•	•		•	•	•
	Links to money	Money Matters		•		•												
	Physical Education	Sports Update				•		•										
	Design & Technology	Construct	•		•	•			•			•	•		•	•		
	Art & Design / Music	The Arts Roundup				•				•							•	
	History	In the Past / Famous Mathematicians			•							•			•	•		
	Geography	Around the World			•	•		•		•				•		•	•	•
	Computing	Technology Today										•						
	Science	Focus on Science			•		•	•	•		•			•			•	•
	English	The Language of Maths	•	•				•		•	•		•	•		•	•
Mathematics		Let's Investigate	•	•		•	•	•	•		•				•		•	•	•
		The Puzzler	•	•	•		•	•	•	•				•	•	•			
		Looking for Patterns				•		•	•			•	•	•				
		What's the Problem?	•		•	•	•	•	•		•	•						

Domain(s)	Topic	Issue
Number: – Fractions	Fractions	17
	Fractions	18
	Fractions	19
	Fractions	20
Measurement	Length	21
	Mass	22
	Capacity and volume	23
	Time	24
	Measurement	25
	Measurement	26
Geometry: – Properties of shapes	2-D shapes	27
	3-D shapes	28
	Symmetry	29
	Position and direction	30
	Movement and angle	31
	Geometry	32
	Geometry	33
Statistics	Statistics	34
	Statistics	35
	Statistics	36

20

Resources used in *Stretch and Challenge 3*

A fundamental skill of mathematics is knowing what resources to use and when it is appropriate to use them. It is for this reason that many of the activities in *Stretch and Challenge 3* give no indication to the children as to which resources to use. Although the teacher's notes that accompany each activity include a list of resources, children should be encouraged to work out for themselves what they will need to use to successfully complete an activity.

It is assumed that for each activity children will have ready access to pencil and paper, and any other resources that are specifically mentioned in an activity, for example, computers or other Resource sheets. However, all other equipment should be left for the children to locate and use as and when they see is appropriate.

A list of all the resources children are likely to need in *Stretch and Challenge 3* is given below.

Resource	1	2	3	4	5	6	7	8	9	10	11	12	13	14	15	16	17	18
pencil and paper	•	•	•	•	•	•	•	•	•	•	•	•	•	•	•	•	•	•
Resource sheets	2, 3	2, 3	2, 3	2, 3	2, 3	2, 3, 9	2, 3, 9	2, 3	2, 3, 9	2, 3	2, 3	2, 3	2, 3	2, 3	2, 3	2, 3	2, 3, 8, 9	2, 3
computer with Internet access		•											•		•	•		
ruler					•	•		•									•	•
scissors								•										
calculator					•	•			•	•	•						•	•
counters	•	•									•	•			•	•		
interlocking cubes																	•	
tape measure																	•	
coloured pencils																	•	
A4 paper																		•
large sheets of paper, i.e. A3 or A2 paper																		•
selection of different take-away menus			•															
art paper and colouring materials			•															
selection of different newspapers and magazines				•														
set of dominoes					•	•		•										
set of 0–9 digit cards								•					•					
1–6 dice								•										
matchsticks													•					
selection of coins (real or play) or counters that represent £1														•				
blank cards																•		

Resource	19	20	21	22	23	24	25	26	27	28	29	30	31	32	33	34	35	36
pencil and paper	•	•	•	•	•	•	•	•	•	•	•	•	•	•	•	•	•	•
Resource sheets	2, 3	2, 3, 8	2, 3	2, 3	2, 3	2, 3	2, 3	2, 3, 8	2, 3, 8	2, 3	2, 3, 8, 9, 12, 13	2, 3, 4, 8, 9, 10, 12, 13	2, 3, 8, 9, 11	2, 3, 8, 9, 12	2, 3, 6, 8, 9, 13	2, 3, 7, 8, 9	2, 3	2, 3, 8, 9	2, 3, 8, 9
computer									•		•					•		
computer with Internet access			•		•	•		•								•		•
data handling software																	•	•
ruler			•				•	•	•	•	•	•	•	•	•		•	•
scissors									•	•	•			•	•			
calculator	•	•		•														
counters							•				•							

21

Introduction

Resource	\multicolumn{18}{c}{Stretch and Challenge Issue}
	19	20	21	22	23	24	25	26	27	28	29	30	31	32	33	34	35	36
A4 paper					•		•		•									
geometric shapes (circle, triangle, square and rectangle) of different sizes									•									
matchsticks									•									
geoboards and elastic bands									•									
art straws									•									
modelling clay				•					•									
sticky tape					•					•								
glue										•	•							
junk construction material										•								
blank dice (or white cubes), red and blue washable felt-tip pens										•								
interlocking cubes							•			•	•			•				
coloured pencils		•									•	•					•	
camera											•							
range of different measuring equipment											•							
"small treasure"												•						
four draughts pieces and four identical chess pieces, e.g. pawns												•						
set of 1–20 number cards														•				
large sheet of coloured square paper															•			
atlas															•			
collection of clothing manufactured in different countries with labels showing country of production			•															
art paper and colouring materials			•													•		
tape measure			•				•	•										
collection of different-sized weights, including 1 kg, 500 g, 100 g and 50 g weights				•														
scale balance				•														
weighing scales				•	•													
marbles, lead beads or similar				•														
Compare Bears or similar				•														
counters, interlocking cubes, beads, crayons, pencils, dice, etc				•														
1-litre, 3-litre and 4-litre jugs					•													
access to water					•													
small counters or cubes					•													
rice, pasta or sand					•													
several sheets of A4 paper (preferably used sheets)							•											
pack of playing cards							•											•
selection of different envelopes							•											
selection of different newspapers							•	•										
metre rule and trundle wheel							•	•										
dictionary (preferably mathematical)																	•	
set of 0–9 digit cards		•																

Issue 1 – Number

S&C Volume 3

The Maths Herald

Name:

Date:

Let's Investigate

Look at this abacus.
It shows the number 536 using 14 beads.
Using all 14 beads, what is the largest number you can make on an abacus?
What is the smallest number you can make on an abacus using 14 beads?

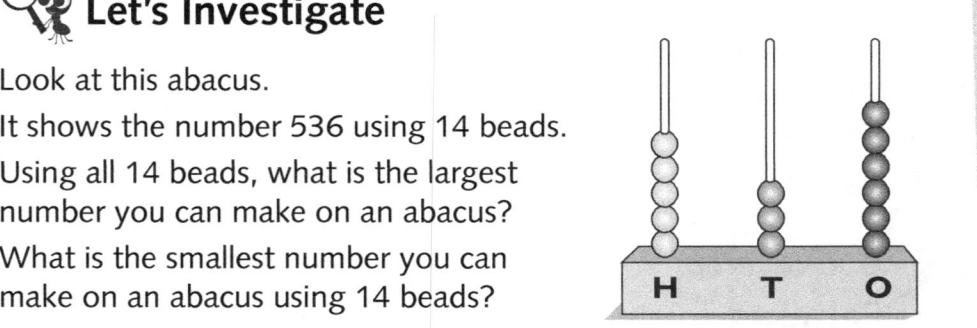

The Puzzler

Rearrange all the digits in each of the black boxes to make a different number. Write this number in the white boxes to the right ▶ and / or below ▼.
One has been done for you.

Issue 1 – Number

What's the Problem?

Mrs and Mrs Keft invite 3 other couples for dinner.
Everyone shakes hands with everyone else, except each husband and wife – they do not shake each other's hands.
How many handshakes occur at dinner?
Write about how you worked out the answer.

What's the Problem?

Use the clues to find the number in the box.

Clue 1: It is a 3-digit number.
Clue 2: All the digits are different.
Clue 3: 2 of the digits are even, 1 is odd.
Clue 4: The sum of all the digits is 13.
Clue 5: All the digits are less than 7.
Clue 6: The tens digit is odd.
Clue 7: The number is greater than 500.

Money Matters

Luis has saved 3 times as much pocket money as Marcus.
Both children have saved a whole but odd number of pounds.
The combined total of their pocket money is more than £20 but less than £30.
How much pocket money does each child have?

© HarperCollins Publishers 2016

Looking for Patterns

This printer's wheel makes this pattern: ▲■■■▲▲...

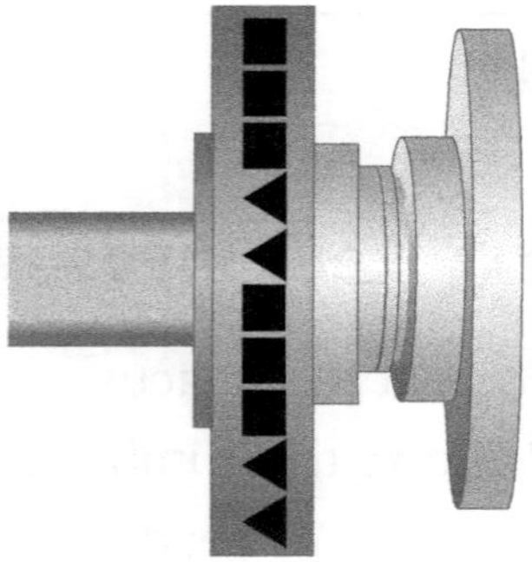

- What will the 40th shape be?
- What position in the pattern will the 46th shape be?
- What shape will be to the right of it?
- What will the 83rd shape be?
- What position in the pattern will the 31st triangle be?

Looking for Patterns

Make up some printer's wheel puzzles similar to the one in the Looking for Patterns activity above.

Ask different questions each time.

Swap your pattern and questions with a friend and answer each other's questions.

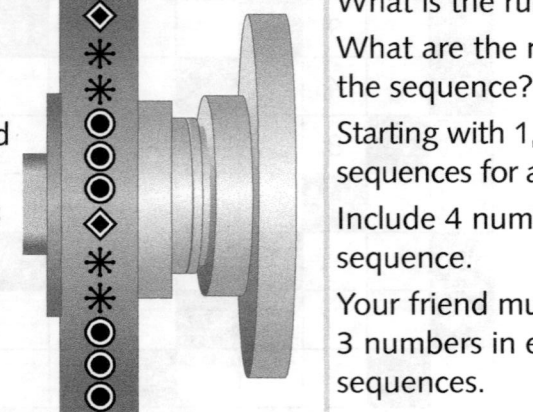

Looking for Patterns

Look at this simple number sequence:

1, 5, 9, 13, …

What is the rule?

What are the next 3 numbers in the sequence?

Starting with 1, write some number sequences for a friend to continue. Include 4 numbers in your sequence.

Your friend must write the next 3 numbers in each of your sequences.

The Puzzler

| 112 | 84 | 136 | 224 | 1182 | 336 |

Only 1 of the numbers above fits all the descriptions below. Can you find the number?

- The sum of all my digits is a multiple of 4.
- I am a multiple of 7.
- My ones digit is twice my tens digit.
- I am a multiple of 8.
- I am more than half of 300.
- I am not a multiple of 3.

Now make up a similar puzzle for a friend to solve.

Let's Investigate

How many pencils do you think there are in your classroom?

Don't count – estimate.

How could you find out without counting them all?

How many pencils do you think there are in your school?

How could you estimate this?

Plan how you could estimate this.

Looking for Patterns

30, 80, 180, 330, …

What is the rule for the number sequence?

Write the next 4 numbers in the sequence.

Issue 2 – Number
S&C Volume 3

The Maths Herald

Name: Date:

The Puzzler

From which box will ticket number 76 fall?

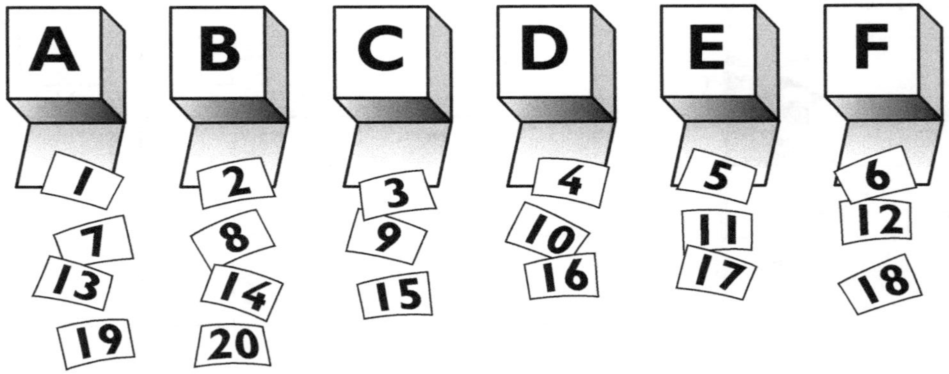

A B C D E F
1 2 3 4 5 6
7 8 9 10 11 12
13 14 15 16 17 18
19 20

Write about how you worked out the answer.

Looking for Patterns

Rasda Supermarket stacks cans of soup in a tower so that the number of cans in each row is one less than the number of cans in the row below it.

If there are 36 cans on display, how many rows of cans are there?

If there were 12 rows of cans, how many cans would be on display?

Write about how you worked out the answers.

© HarperCollinsPublishers 2016

Issue 2 – Number

Looking for Patterns

12 elves are standing in a circle.

Every 3rd elf is wearing pointy shoes and every 4th elf is wearing a green cap.

How many elves are not wearing either pointy shoes or green caps?

Let's Investigate

Approximately how many buttons are there in your classroom?

Write about how you arrived at your estimate.

Sports Update

Morcella Rovers and Cerqueto Rangers played each other twice during the season.

Both matches were a draw.

The final score of the first match was 2–2 and the final score of the second match was 4–4.

What are the different possible half-time scores for each match?

What do you notice about the total number of possible scores for each match?

What do both these numbers have in common?

Issue 2 – Number

The Puzzler

Write each of the digits 1 to 5 once only in each row and column.

Use the less than (<) and greater than (>) signs to help you.

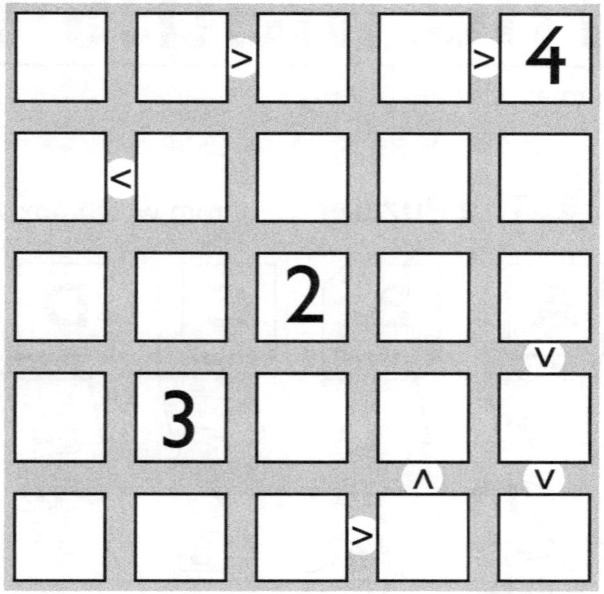

```
[ ] > [ ] [ ] < [ ] [ ]
[ ] [ ] [ ] [ ] [ ]
[ ] [3] [2] [ ] [4]
[ ] > [ ] > [ ] [ ] [ ]
[ ] [ ] [ ] < [ ] < [ ]
```

Looking for Patterns

This is a 1–100 number square, but not as you know it!

Continue the pattern of the numbers to find the number that belongs in the ? square.

Can you find ? without numbering all the blank squares?

What patterns do you notice?

1	2	3	4	5	6	7	8	9	10
36	37	38	39	40	41	42	43	44	11
35	64	65	66	67	68	69	70	45	12
34	63						71	46	13
33	62			?			72	47	14
32	61						73	48	15
31	60							49	16
30	59							50	17
29	58	57	56	55	54	53	52	51	18
28	27	26	25	24	23	22	21	20	19

Looking for Patterns

The number on each guestroom door in The Grand Hotel is written as a 3-digit number. The first digit stands for the floor number, and the rest of the number is the guestroom number.

For example:
- 101 — This is Room 1 on the 1st floor.
- 410 — This is Room 10 on the 4th floor.

The Grand Hotel has a total of 8 guestroom floors from the 1st floor to the 8th floor, with 15 guestrooms on each floor.

How many guestroom doors have the number 7 on them?

How many guestroom doors have a zero on them?

The Language of Maths

What is the …

- largest 4-digit number that has 2 as one of its digits?
- smallest 4-digit number that has 2 as one of its digits?
- largest 4-digit number in which no digit occurs more than once?
- smallest 4-digit number in which no digit occurs more than once?
- largest 4-digit number that has a 2 and in which no digit occurs more than once?
- smallest 4-digit number that has a 2 and in which no digit occurs more than once?

© HarperCollins*Publishers* 2016

Issue 3 – Number
S&C Volume 3

The Maths Herald

Name:
Date:

Construct

Work with a friend to design and create your own take-away menu.

Things to think about:
- What type of food are you going to sell?
- What will be on your menu, and what prices are you going to charge?
- What will your opening hours be?
- Will you be offering a home delivery service? If so, what will you charge?

Make sure that your menu has all the relevant information you need to operate a successful business.

Let's Investigate

Investigate take-away menus.
- What do they all have in common? Make a list of all the common features that different take-away menus share.
- What makes a good take-away menu?
- Compare all the take-away menus you have. Which menus offer good value for money? Which ones are expensive? Justify your conclusions.

© HarperCollins*Publishers* 2016

The Language of Maths

- Which numbers fit several of the statements?
- Which numbers fit only 1 statement?
- Write a different statement that fits 2 of the numbers.
- Write a different statement that fits 5 of the numbers.
- For each statement write another 2 numbers.

Numbers
–6
25
2·75
50%
65 042
1010
$14\frac{1}{2}$
–14
48
$\frac{3}{4}$

Statements
mixed number
between 0 and 100
square number
more than 1000
multiple of 5
percent
greater than –10
less than 2
odd number
negative number

The Arts Roundup

Use the following clues to find out how many seats each of these theatres hold.

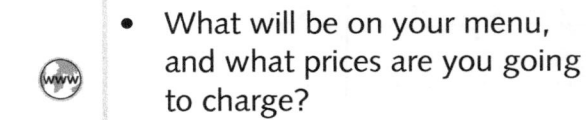

Clue 1: The Studio has half as many seats as The Actor's Space.

Clue 2: The two largest theatres have a royal connection.

Clue 3: The number of seats in The Queen's Centre is not a multiple of 5.

Theatre
The Palace Theatre
The Actor's Space
The Playhouse
The Queen's Centre
The Studio

Number of seats
157
286
314
528
635

Issue 3 – Number

Looking for Patterns

A school is raising money by selling tickets for a tombola. They sell 140 tickets numbered 1 to 140.

How many of the tickets have the number 3 on them?

How many times does the number 3 appear on the 140 tickets?

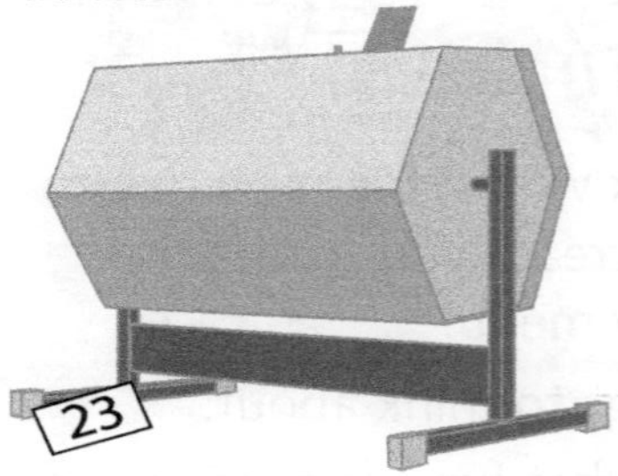

Let's Investigate

Approximately how many books are there in your school? (Do not count exercise books.)
Write about how you arrived at your approximation.
Approximately how many exercise books are there in your school?
Write about how you arrived at your approximation.
Which approximation do you think is closer to the actual number of books? Why?

Sports Update

Every Saturday morning between 9:00 a.m. and 10:00 a.m., Cindy Longlegs runs an exercise class. Last Saturday there were a total of 20 people in the class.
For every 3 women, there was 1 man.
How many women were in the class?
How many men were there?
Write about how you worked out the answer.

Issue 3 – Number

The Puzzler

Insert the greater than or less than signs in the boxes to make each statement correct, for example: 5 > 3 < 4

9 ☐ 4 ☐ 7 ☐ 3 ☐ 5   17 ☐ 2 ☐ 1 ☐ 8 ☐ 7

7 ☐ 11 ☐ 10 ☐ 8 ☐ 12   11 ☐ 12 ☐ 4 ☐ 3 ☐ 6

2 ☐ 3 ☐ 14 ☐ 6 ☐ 7   8 ☐ 7 ☐ 6 ☐ 10 ☐ 14

Around the World

Most weather forecasts give the expected daily high (maximum) and low (minimum) temperature readings.

Search the internet for the weather forecast in different parts of the world. Try and locate places where it is wintertime.

Write down the following:
- name of city or town
- forecast maximum temperature
- forecast minimum temperature

Order the cities by their maximum temperatures, from lowest maximum temperature to highest maximum temperature.

Order the cities by their minimum temperatures, from lowest minimum temperature to highest minimum temperature.

Are the orders of the cities the same in both lists? Why do you think this is?

© HarperCollins*Publishers* 2016

Issue 4 – Number
S&C Volume 3

The Maths Herald

Name: Date:

What's the Problem?

There are 9 competitors in a spelling competition, numbered 1 to 9.

A friend asks one of them what their number is.

Their friend replies:

"If the number of numbers less than mine is multiplied by the number of numbers greater than mine, the product would be the same as it would be if my number was 2 greater than it is."

What is the competitor's number?

Spell-a-thon

Technology Today

The following phone numbers are actually codes. Complete the letters on the phone keypad and use the letters to decode each of the following:

season: 78 6637   fruit: 277 4268
day: 7288 7329   vegetable: 76 8286
month: 6683 6237   city: 54 837 7665
planet: 637 8863   country: 8424 5263

Encode a city, country and a musical instrument for a friend to decode.

© HarperCollins*Publishers* 2016

Let's Investigate

The number 123 is a *3-digit number*. It is made up of three digits: 1, 2 and 3.

The sum of the 3 digits in the number 123 is 6 because 1 + 2 + 3 = 6.

What are all the other 3-digit numbers whose digits total 6?

Write about how you found all the numbers possible.

The Language of Maths

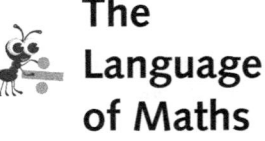

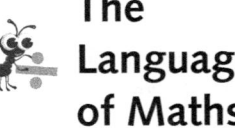

There are 2 tribes on the island of Neb.

The Yays always tell the truth and the Nays always lie.

You meet 2 people.

The first says, "Both of us are Nays."

Which tribe do each of them belong to?

The Language of Maths

F, M, A, M, J, J, A, ...

o, t, t, f, f, s, s, ...

What are the next 3 letters in each of these sequences?
What is the rule for each sequence?

Issue 4 – Number

Famous Mathematicians

Diophantus of Alexandria was a Greek mathematician who was interested in patterns in numbers. He discovered an interesting pattern involving *triangular numbers*.
A triangular number is a number that can be drawn as a triangular grid of dots, where the first row contains one dot and each following row contains one more dot than the previous row.

- Multiply each triangular number by 8 and add 1. What do you notice about these numbers? Diophantus was the first to discover this pattern nearly 2000 years ago!

decimal

What's the Problem?

In early spring, Greenfingers Nursery planted 120 tulip bulbs. Out of every 6 bulbs they planted, 5 bulbs grew into tulips. How many of the bulbs grew into tulips?

Issue 4 – Number

Let's Investigate

Look through a selection of newspapers and magazines and find as many different decimals as you can. Cut them out and arrange them in order, smallest to largest.
Can you explain what all the decimals mean?

The Puzzler

What number is on each of the cards?

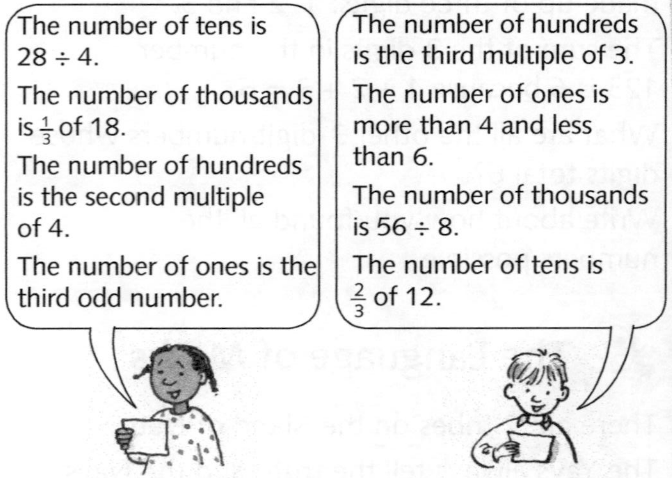

- The number of tens is $28 \div 4$.
- The number of thousands is $\frac{1}{3}$ of 18.
- The number of hundreds is the second multiple of 4.
- The number of ones is the third odd number.

- The number of hundreds is the third multiple of 3.
- The number of ones is more than 4 and less than 6.
- The number of thousands is $56 \div 8$.
- The number of tens is $\frac{2}{3}$ of 12.

Secretly choose a number and make up clues for the number. Give your clues to a friend so that they can work out your number.

Focus on Science

Sunita's class went on a visit to the zoo. In one display there were a group of eight-legged spiders, and in another display a set of six-legged beetles. Altogether Sunita counted 10 beetles and spiders, and 74 legs.
How many beetles were on display? How many spiders were there?

© HarperCollins*Publishers* 2016

The Maths Herald

S&C Volume 3
Issue 5 – Addition

Name: _____ Date: _____

Let's Investigate

| | 1 | 2 | 3 | 4 | 5 |

Choose 3 of the brick numbers from above.

Write the numbers, in any order you like, on the bottom row of the wall.

Then add pairs of adjacent bricks to find the numbers for the middle row.

Then add the 2 numbers in the middle row to give the number for the top brick.

By rearranging the same 3 numbers on the bottom row, how many different numbers can you get for the top brick?

What happens if you choose 3 different bricks for the bottom row?

The Puzzler

Write the digits 1 to 9 on the 3 × 3 grid so that each row and column of 3 digits totals the numbers in the circles.

			⑨
			⑰
			⑲
⑫	⑬	⑳	

© HarperCollins*Publishers* 2016

Let's Investigate

A *palindrome* is a word, number or phrase that reads the same backwards as forwards. For example:

MADAM   NURSES RUN   5775   2002

The numbers 11, 22, 33, … 99 are all examples of 2-digit palindromic numbers.

You can make any 2-digit number into a palindromic number using these steps.

Step ①: Start with a 2-digit number
Step ②: Reverse the digits
Step ③: Add the 2 numbers together.

Repeat Steps ② and ③ until the number becomes palindromic.

Look at these examples:

Example 1
① 43
② 34
③ 43 + 34 = 77

The number 43 becomes palindromic after 1 stage.

Example 2
① 62
② 26
③ 62 + 26 = 88

The number 62 becomes palindromic after 1 stage.

Example 3
① 58
② 85
③ 58 + 85 = 143
① 143
② 341
③ 143 + 341 = 484

The number 58 becomes palindromic after 2 stages.

Example 4
① 59
② 95
③ 59 + 95 = 154
154
② 451
③ 154 + 451 = 605
605
② 506
③ 605 + 506 = 1111

The number 59 becomes palindromic after 3 stages.

Investigate which 2-digit numbers become palindromic in 2, 3, 4, 5 or 6 stages.

There are two 2-digit numbers that do not become palindromic until after 24 stages! What are these 2 numbers?

Issue 5 – Addition

Let's Investigate

Children in Year 1 learn by heart all the addition number facts with answers up to, and including, 20.

On the right are all the addition number facts for 5.

What are all the addition number facts to 10?

Altogether, how many addition number facts to 10 do Year 1 children need to learn by heart?

Let's Investigate

This addition calculation can be represented using a set of normal dominoes.

$$\begin{array}{r} 31 \\ +123 \\ \hline 154 \end{array}$$

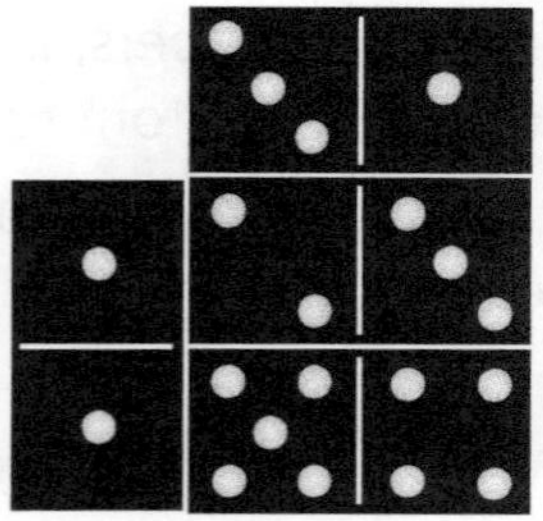

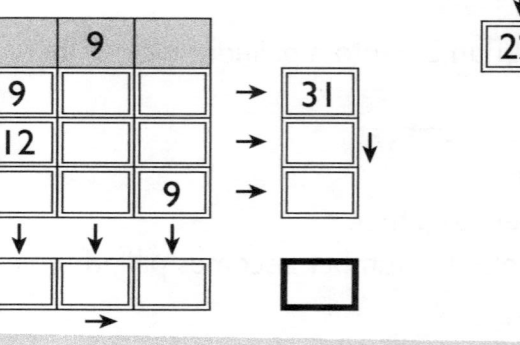

What other addition calculations can you make using dominoes?

Looking for Patterns

Complete these addition calculations.

Following the same pattern as these calculations:

- What 3 odd numbers have a total of 75?
- What 3 even numbers have a total of 102?

Write about what you did.

$1 + 3 + 5 = \bigcirc$

$3 + 5 + 7 = \bigcirc$

$5 + 7 + 9 = \bigcirc$

$7 + 9 + 11 = \bigcirc$

$2 + 4 + 6 = \bigcirc$

$4 + 6 + 8 = \bigcirc$

$6 + 8 + 10 = \bigcirc$

$8 + 10 + 12 = \bigcirc$

The Puzzler

Look carefully at the completed grid on the right.

Use the same rules to complete the grid below.

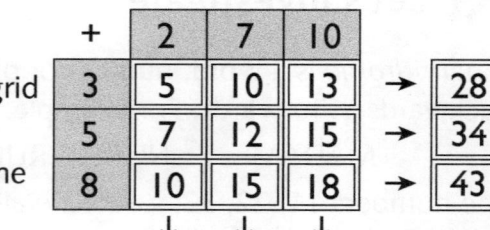

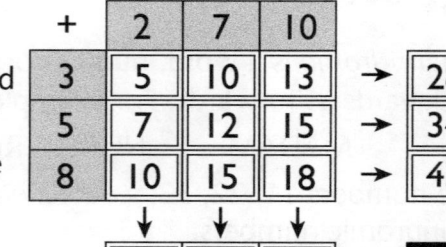

+			
5	9		
	12		
		9	

→ 31

+	2	7	10	28
3	5	10	13	34
5	7	12	15	43
8	10	15	18	
	22	37	46	105

commutative law

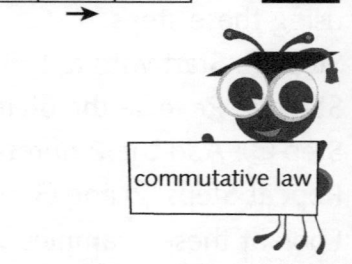

The Puzzler

Fabio and Gabby are playing a game using 4 sticky balls and this board.

Fabio threw the 4 balls. His score was 146. What numbers did the balls hit?

Gabby took her turn. She scored 40 more points than Fabio.

What 4 numbers did Gabby hit?

(spinner numbers: 29, 16, 57, 80, 63, 41, 32, 78)

Issue 6 – Addition
S&C Volume 3

The Maths Herald

Name: Date:

The Puzzler

AAH + AAH = HARP

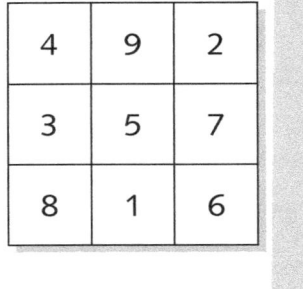

Each of the letters stands for a different digit.
Which digit does each stand for?
Explain how you worked out what each letter stands for.

Money Matters

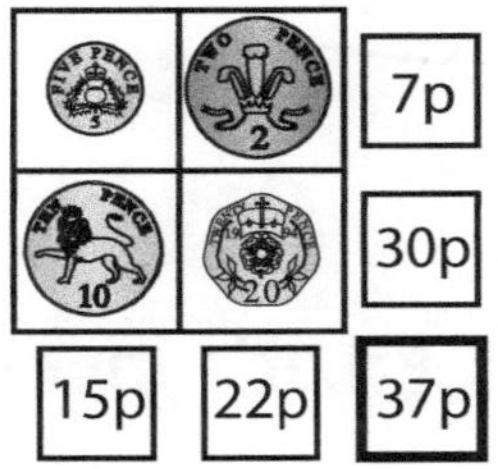

Look at the puzzle on the right.
Apply the same rules to the 2 puzzles below to fill in the missing coins and totals.

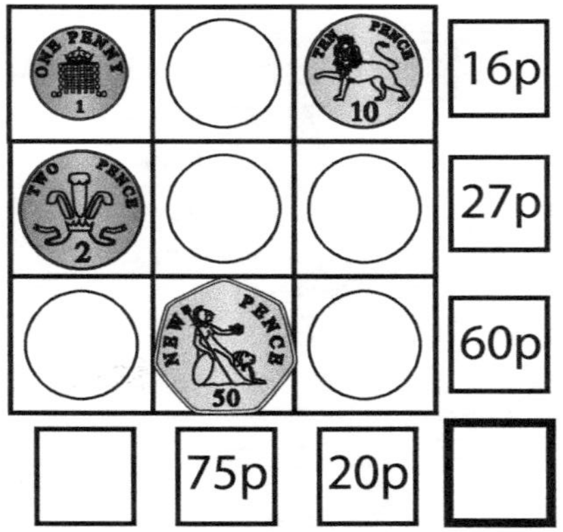

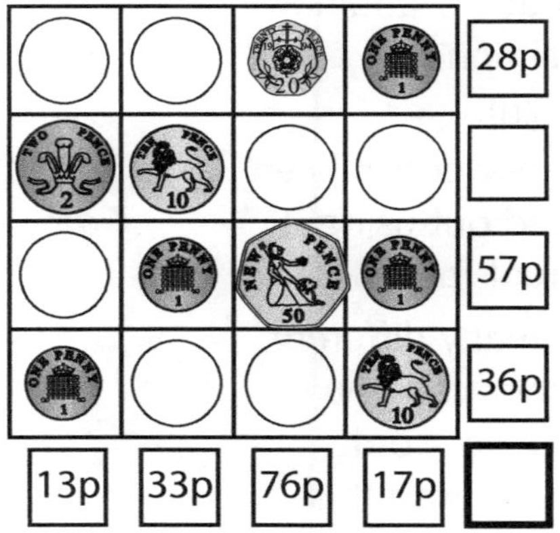

© HarperCollinsPublishers 2016

Issue 6 – Addition

Let's Investigate

This is a 3 × 3 magic square.

4	9	2
3	5	7
8	1	6

The sum of each column, row and diagonal is the same – this is the magic number.

What is the magic number for this magic square?

Is the square still magic if you:
① add 2 to each number?
② subtract 1 from each number?
③ multiply each number by 2?
④ multiply each number by 10?
⑤ add ½ to each number?
⑥ subtract ½ from each number?

Look at this 4 × 4 magic square.

4	14	15	1
9	7	6	12
5	11	10	8
16	2	3	13

What is the magic number for this magic square?

Is the square still magic if you make changes ① to ⑥ above?

Looking for Patterns

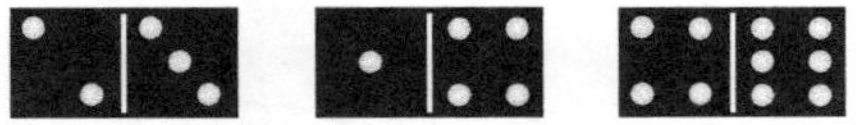

The 3 dominoes above have been arranged to make the addition calculation on the right.

```
  23
+ 41
----
  64
```

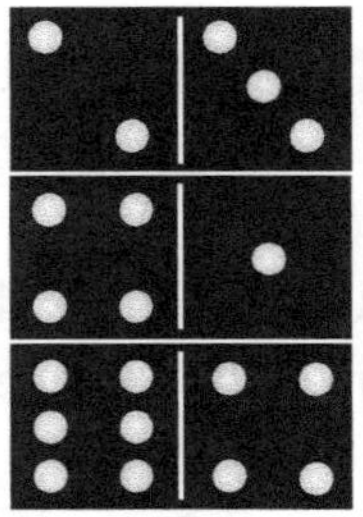

Arrange these 9 dominoes to make 3 similar addition calculations.

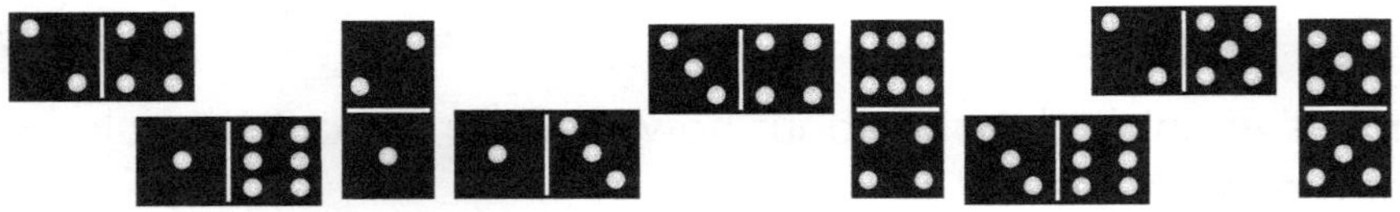

Issue 6 – Addition

Money Matters

You have exactly £5 to spend in the Joke Shop.
What things can you buy?
How many different combinations of things can you buy?

- Invisible Ink 95p
- Whoopee 70p
- Soapy Sweets 80p
- 50p (eyes)
- Joke Book £1
- Ladybird 45p
- Party String 60p
- Magic Tricks £1.50

Let's Investigate

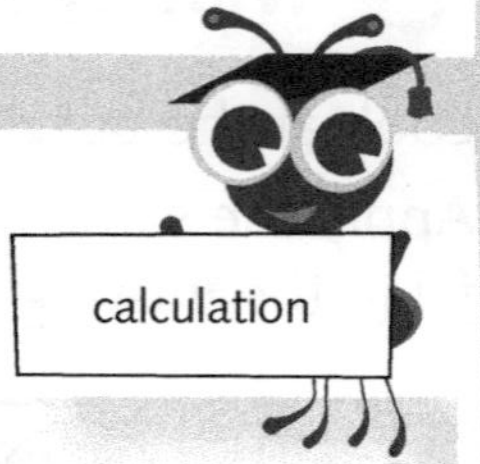

Look at the addition calculation on the right.
The answer has been worked out using 2 different methods.
How many different methods can you use to work out the answer to this addition calculation?

284 + 157

calculation:

85 + 37
= 85 + 30 + 7
= 115 + 7
= 122

85 + 37
= 80 + 30 + 5 + 7
= 100 + 12
= 122

The Puzzler

In each of the 4 puzzles below, the number in the star is made by adding the 2 numbers at the corners.
Write the missing numbers in the circles and stars.

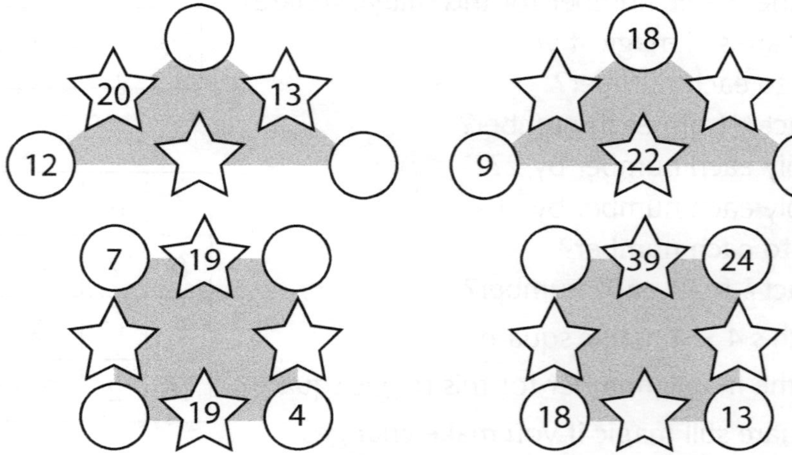

(puzzles with stars showing 20, 7, 19, 19, 13, 12, 4 and 18, 22, 39, 24, 13, 9)

Write a triangle and a square addition puzzle for a friend to solve.

Looking for Patterns

Use each of the digits 0 to 9 only once to complete these addition number sentences.

0, 1, 2, 3, 4, 5, 6, 7, 8, 9

3☐ + ☐6 = 4☐   2☐ + ☐9 = 74

☐7 + 25 = ☐2   7☐ + 34 = 1☐☐

When you've finished, write about how you worked out the value of each of the missing digits.

The Maths Herald

Issue 7 – Subtraction
S&C Volume 3

Name: _____ Date: _____

🐞 Let's Investigate

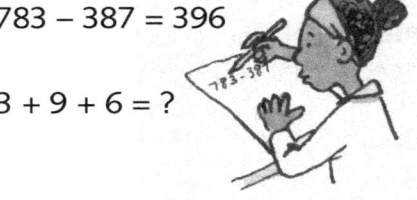

Take any 3-digit number.   387
Reverse the digits.   783
Find the difference between the   783 – 387 = 396
2 numbers.
Add the digits of the answer together. 3 + 9 + 6 = ?
Repeat for other 3-digit numbers.
What do you notice?

🐞 Let's Investigate

Look at this subtraction calculation.
```
   386
 - 178
 -----
```

What is the answer?
Write about your method for working it out.

Look at this method for working it out.
Can you explain what is happening?

```
300 (+ 86)         86
- 178            + 122
-----            -----
  122             208
                    1
```

```
 835      359       773
-347     -168      -684
----     ----      ----
```

```
 573      743
-495     -256
----     ----
```

Is this a method you might use? If not, why not? If so, when?

© HarperCollinsPublishers 2016

🐞 Let's Investigate

Using only the digits 7, 8 and 9, how many different "2-digit subtract 2-digit" calculations can you make with answers of 9, 10, 11 and 12?

You can use each of the 3 digits more than once in a calculation.

☐☐ − ☐☐ = 9   ☐☐ − ☐☐ = 10

☐☐ − ☐☐ = 11   ☐☐ − ☐☐ = 12

🐞 Let's Investigate

As well as differences of 9, 10, 11 and 12, you can use the digits 7, 8 and 9 to make other "2-digit subtract 2-digit" calculations that have other differences.

You can use each of the 3 digits more than once in a calculation.

☐☐ − ☐☐ = ?

| Don't include calculations with a difference of 0. That's just too easy! |
| 77 − 77 = 0 |

🐜 The Puzzler

Each letter in the word MINUS stands for a different digit.

Work out what digit each of the letters stands for.

Issue 7 — Subtraction

Let's Investigate

Children in Year 1 learn by heart all the subtraction number facts with answers up to, and including, 20.

On the right are all the subtraction number facts for 5.

What are all the subtraction number facts to 10?

Altogether, how many subtraction numbers facts do Year 1 children need to learn by heart?

5 − 0 = 5
5 − 1 = 4
5 − 2 = 3
5 − 3 = 2
5 − 4 = 1
5 − 5 = 0

Let's Investigate

Look at the subtraction calculation on the right.
The answer has been worked out using 2 different methods.

inverse relationship

63 − 26
= 63 − 20 − 6
= 43 − 6
= 37

37 ← −3 ← 40 ← −3 ← 43 ← −20 ← 63

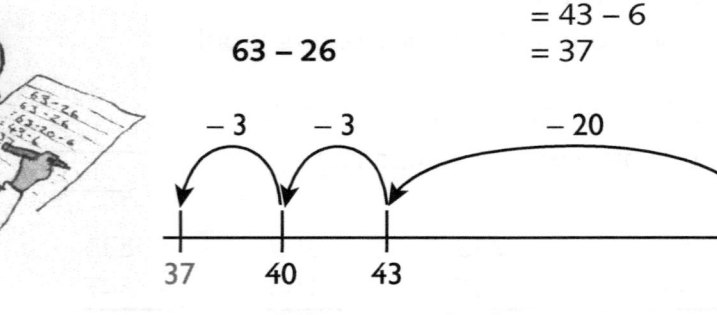

How many different methods can you use to work out the answer to this subtraction calculation?

264 − 178

The Puzzler

In each of the four puzzles below, the number in the star is the difference between the 2 numbers at the corners.

Write the missing numbers in the circles and stars.

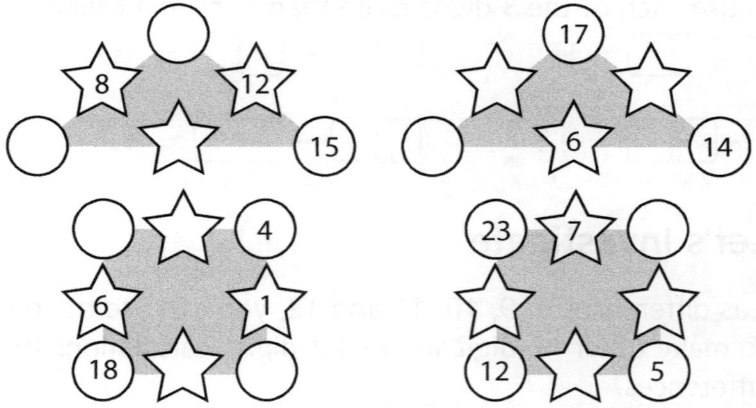

(puzzle with numbers: 8, 12, 15, 6, 18, 4, 1, 23, 7, 14, 17, 12, 5)

Write a triangle and a square subtraction puzzle for a friend to solve.

In the Past

The Coliseum in Rome is the largest Roman amphitheatre in the world. It could seat 50 000 people and was used for gladiator contests and public spectacles like mock battles.

It is oval in shape and is 189 m long and 156 m wide. The central arena is also oval and is 87 m long by 54 m wide. The central arena is surrounded by rows of seats.

How much greater is the Coliseum's length than its width?
How much greater is the central arena's length than its width?
How wide is the seating that runs around the arena?

© HarperCollinsPublishers 2016

Issue 8 – Subtraction
S&C Volume 3

The Maths Herald

Name:

Date:

The Puzzler

ATT – S = SA

Each of the letters stands for a different digit. Which digit does each stand for? Explain how you worked out what each letter stands for.

Let's Investigate

1, 2, 3, 4

Using only 3 of the digits 1, 2, 3 and 4 in each calculation, investigate what numbers you can make by finding the difference between a 2-digit and a 1-digit number.

□□ – □ = ?   prediction

Each digit can only be used once in each calculation.

You can make the numbers 11 and 41 like this:

14 – 3 = 11   42 – 1 = 41

© HarperCollins*Publishers* 2016

The Puzzler

Each of the 3 shapes below can be folded to make a 1–6 dice.

Draw dots on the blank squares so that opposite faces of the dice add up to 7.

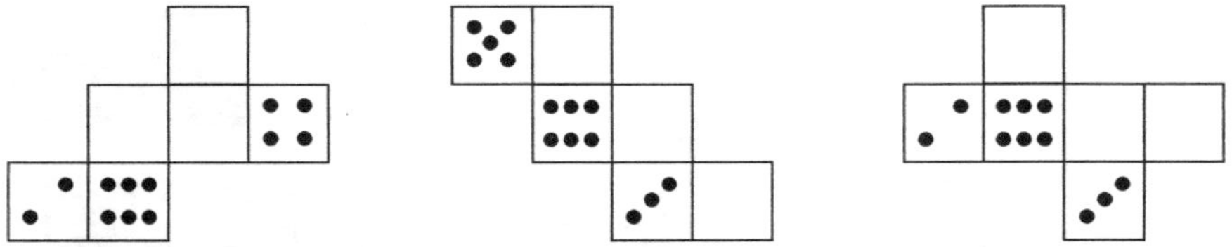

Around the World

This table shows the 28 member states of the European Union (EU), along with their capital cities and the capital cities' populations.

Write different statements comparing the populations of the different capital cities of the EU. Here are some to get you started.

- 800 000 more people live in Stockholm than in Luxembourg.
- There are 400 000 fewer residents in Vienna than Paris.

Country	Capital city	Capital city population
Austria	Vienna	1 800 000
Belgium	Brussels	1 200 000
Bulgaria	Sofia	1 200 000
Croatia	Zagreb	790 000
Cyprus	Nicosia	250 000
Czech Republic	Prague	1 200 000
Denmark	Copenhagen	1 200 000
Estonia	Tallinn	430 000
Finland	Helsinki	620 000
France	Paris	2 200 000
Germany	Berlin	3 500 000
Greece	Athens	3 000 000
Hungary	Budapest	1 700 000
Ireland	Dublin	520 000
Italy	Rome	2 800 000
Latvia	Riga	640 000
Lithuania	Vilnius	530 000
Luxembourg	Luxembourg	100 000
Malta	Valletta	6400
Netherlands	Amsterdam	800 000
Poland	Warsaw	1 700 000
Portugal	Lisbon	540 000
Romania	Bucharest	1 900 000
Slovakia	Bratislava	410 000
Slovenia	Ljubljana	270 000
Spain	Madrid	3 100 000
Sweden	Stockholm	900 000
United Kingdom	London	8 600 000

Let's Investigate

On the wall below there are 3 pairs of tiles, side-by-side or one above the other, where the 2 numbers have a difference of 8.

1	18	5	14	11
7	12	22	2	23
10	0	16	8	19
6	20	24	13	4
17	3	9	21	15

Here they are:

16 | 8   16 / 24   13 / 21

There are 5 pairs of tiles where the 2 numbers have a difference of 6.

Can you find them?

Now find pairs of numbers with a difference of 3, 4 and 5.

What's the Problem?

Lucy bought a goldfish and took it home in a plastic bag. Unfortunately the bag had a small leak in it and leaked 12 ml each minute.

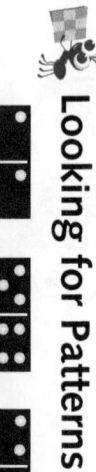

The bag contained 1 litre of water when she left the shop and her journey home took 35 minutes.

Did Lucy get the goldfish home safely? Explain your answer.

Money Matters

Orson had a hole in the pocket in which he kept his money.

On his way to school he lost £1.35.

During the morning he lost another 45p. He spent £2.40 on lunch, lost another 35p in school during the afternoon, and lost 92p on the way home.

When he got home he found that he had exactly half as much money as he had had when he set off for school. How much money did he set off to school with?

Looking for Patterns

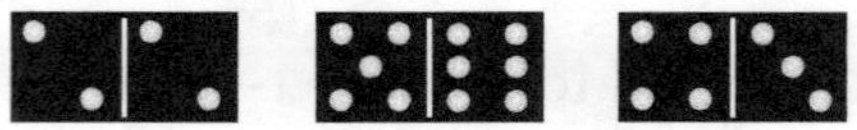

The 3 dominoes above have been arranged to make the subtraction calculation on the right.

$$\begin{array}{r} 56 \\ -\ 34 \\ \hline 22 \end{array}$$

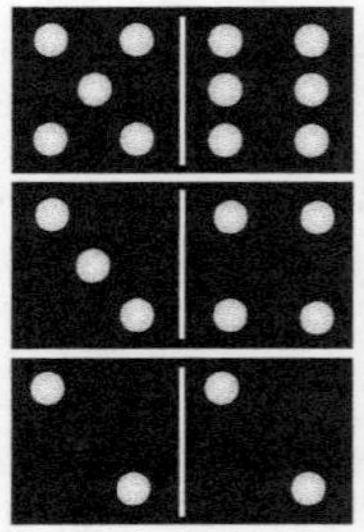

Arrange these 9 dominoes to make 3 similar subtraction calculations.

Let's Investigate

Use 6 1–9 digit cards to make "2-digit subtract 2-digit" calculations, including the answer.

Here is one: **9 8 – 2 5 = 7 3**

How many different calculations can you make?

What's the Problem?

Conroy has 2 old analogue clocks, neither of which keeps the correct time.

One of them loses 1 hour in every 12 hours, and the other loses 2 hours every 12 hours.

If they both start off showing the same time, how long will it be before they do so again?

© HarperCollins*Publishers* 2016

Issue 9 – Multiplication
S&C Volume 3

The Maths Herald

Name:

Date:

The Puzzler

4 8 9 12 16 18 21 24 35 40 45

Write the digits 1 to 9 on the 3 × 3 grid so that pairs of adjacent digits, both horizontally and vertically, make only the products above.

For example:

6		
3		
	8	3

Product of 18   Product of 24

The Puzzler

3 × CAT = AAA

Each of the letters stands for a different digit.

Which digit does each stand for?

Explain how you worked out what each letter stands for.

Money Matters

The Big Read Bookshop opens its doors at 9 o'clock.

In the first hour they take £10.

In every hour after that, they take twice as much as the hour before.

How much money will they take between 4 o'clock and 5 o'clock?

If the shop closes at 6 o'clock, how much money will they have taken during the entire day?

The Language of Maths

Sinbad found a magic casket. Whatever was left in it overnight would be doubled by the next morning.

Sinbad put a gold coin in the casket and left it alone. After 2 weeks the casket was full of gold coins.

After how long was the casket half full of gold coins?

What's the Problem?

Sinbad has a magic casket. Whatever is left in it overnight is doubled by the next morning.

Sinbad let his friend Ahmed use it, but charged him 6 gold coins each night before he let him put anything into it.

After using it for 6 nights, Ahmed found that he had the same number of gold coins as he started with.

How many gold coins did Ahmed start with?

Let's Investigate

25 × 9

Look at the multiplication calculation on the right.

The answer has been worked out using 2 different methods.

How many different methods can you use to work out the answer to this multiplication calculation? **172 × 4**

```
    20    5
 9│180    45  = 225

     25
   ×  9
   180  (20 × 9)
    45  ( 5 × 9)
   225
     1
```

scaling

© HarperCollins*Publishers* 2016

Issue 9 – Multiplication

The Puzzler

Thinking of all the answers to the multiplication facts for the 3, 4, 5, 8 and 10 multiplication tables, which of the 6 boxes on this diagram do you think will have the most numbers in it?

Which box do you think will have fewest numbers?

Once you've made your predictions, write all the answers to the multiplication facts for the 3, 4, 5, 8 and 10 multiplication tables on this diagram. Were your predictions correct?

Multiple of 3	Multiple of 8
Multiple of 4	Multiple of 10
Multiple of 5	

What's the Problem?

Bridget lives on a farm. This is part of an email she sent to her cousin in New Zealand.

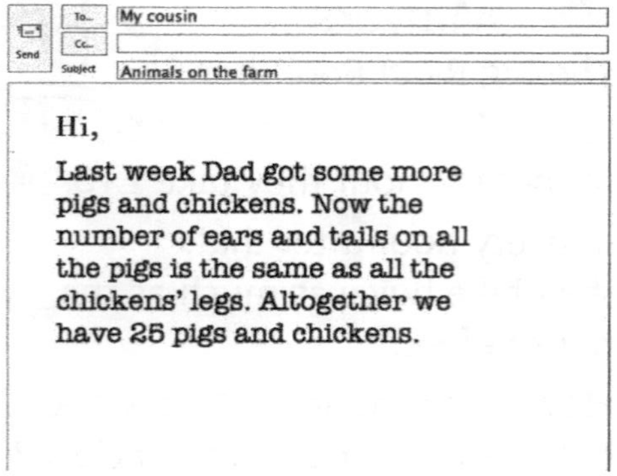

How many sheep are there on the farm?

How many chickens are there?

How did you work out how many of each animal there are?

> Hi,
>
> Last week Dad got some more pigs and chickens. Now the number of ears and tails on all the pigs is the same as all the chickens' legs. Altogether we have 25 pigs and chickens.

Animals on the farm

Let's Investigate

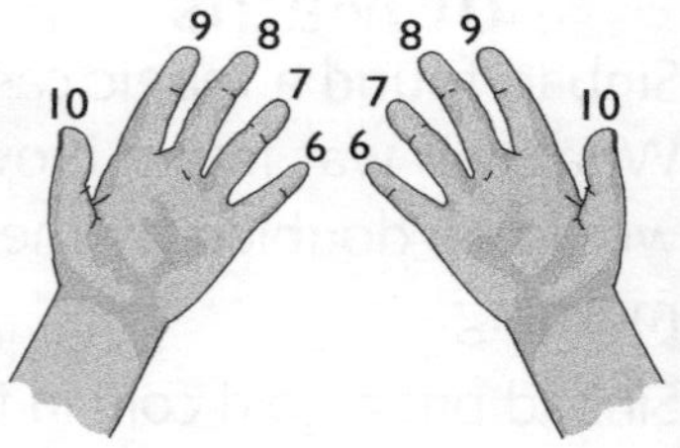

Investigate the following method of working out the multiplication facts from 6 × 6 to 9 × 9, using the fingers of both hands.

To use this method, you must know by heart the 2, 3, 4 and 5 multiplication facts.

Assign the numbers 6 to 10 to the digits of each finger, starting from the little finger (6) to the thumb (10).

Example 1: 6 × 8

- Touch the "6 finger" with the "8 finger".
- Add together the fingers touching and the fingers below the touching fingers (2 + 2 = 4).
- Multiply this total by 10 (4 × 10 = 40).
- Multiply the fingers above the "touch" on your left hand by the fingers above the "touch" on your right hand (4 × 2 = 8).
- Add the two answers together (40 + 8 = 48).
- So, 6 × 8 = 48.

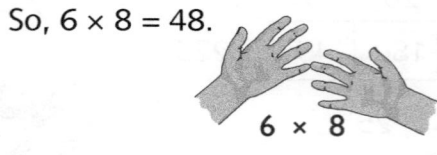

Example 2: 7 × 9

- Touch the "7 finger" with the "9 finger".
- Add together the fingers touching and the fingers below the touching fingers (2 + 4 = 6).
- Multiply this total by 10 (6 × 10 = 60).
- Multiply the fingers above the "touch" on your left hand by the fingers above the "touch" on your right hand (3 × 1 = 3).
- Add the two answers together (60 + 3 = 63).
- So, 7 × 9 = 63.

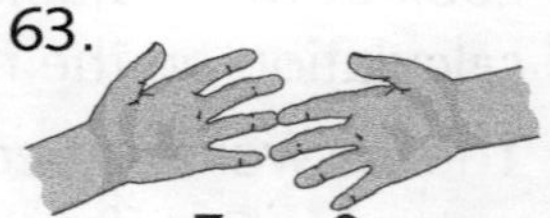

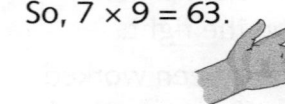

Investigate using this method for other multiplication facts from 6 × 6 to 9 × 9.

Write what you think about this method. What are its strengths? What are its limitations?

© HarperCollins*Publishers* 2016

Issue 10 – Multiplication

S&C Volume 3

The Maths Herald

Name: 　　　　　　　　　　　Date:

The Puzzler

3 × LEG = EEL

Each of the letters stands for a different digit.

Which digit does each stand for?

Explain how you worked out what each letter stands for.

product

Looking for Patterns

Amazing 37

The number 37 is an amazing number. When it is written as part of a multiplication calculation it creates some interesting patterns.

For each set of calculations, identify the pattern and write the next 3 calculations.

Check the answers to your calculations using a calculator.

37 × 3 = 111   37 × 3 = 111
37 × 6 = 222   37 × 33 = 1221
37 × 9 = 333   37 × 333 = 12 321

Now look at the answers to all the calculations.

What do you notice if you add together all the digits in each answer? For example:

37 × 3 = 111 and 1 + 1 + 1 = 3   37 × 33 = 1221 and 1 + 2 + 2 + 1 = 6

Let's Investigate

0, 1, 2, 3, 4, 5, 6, 7, 8, 9

Using the digits 0–9 how many different calculations can you write for each of these number sentences?

You can use the same digit more than once in each calculation.

☐ × ☐☐ = 120   ☐☐ × ☐☐ = 144

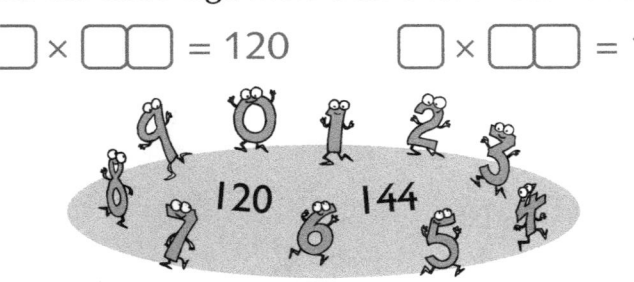

120   144
120   144

Technology Today

Sam is considering changing his mobile phone contract and is looking at different options.

One option costs £10 per month for the first 6 months and then £19 a month for the remaining 18 months.

Another option is £17 per month for 12 months, with the option of renewing it at the same price at the end of the 12 months.

Both contracts offer the same amounts of time, texts and data use each month.

Which option would you advise Sam to take and why?

© HarperCollins Publishers 2016

Issue 10 – Multiplication

In the Past

The Ancient Egyptians did not know how to multiply 2 numbers together to find the product.

Instead, they used a method called duplation, which means doubling numbers over and over again.

For example, to multiply 17 by another number they used the following table:

1 × 17 =	**17**
2 × 17 =	17 × 2 = **34**
4 × 17 =	34 × 2 = **68**
8 × 17 =	68 × 2 = **136**

So, the answer to 5 × 17 is:

17 (1 × 17)
+ 68 (4 × 17)
―――
85
 ¹

So, the answer to 7 × 17 is:

17 (1 × 17)
34 (2 × 17)
+ 68 (4 × 17)
―――
119
  ¹

Use the duplation table above to work out the answers to the following multiplication calculations:

17 × 3   17 × 6   17 × 9   17 × 12

Check your answers using a different method.

Looking for Patterns

Complete the boxes.

3	9
18	36

5	25
50	100

4	16
32	64

	6
	64

	324

The Puzzler

Write numbers in the boxes so when the numbers next to each other are multiplied together they equal the number in the box above.

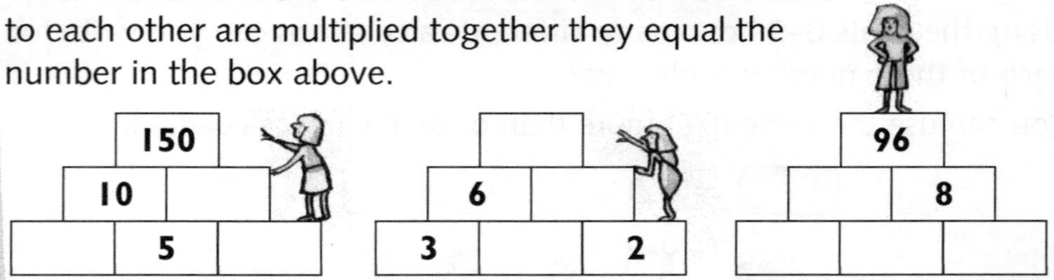

150	
5		3	6	2

96	
	8

Let's Investigate

1, 2, 3, 4

Using only the digits above, what is the largest product you can make for each of these number sentences?

☐ × ☐ = ☐

☐☐ × ☐ = ☐

☐☐ × ☐☐ =

The Puzzler

☐ × ☐ × ☐ = 366

Consecutive numbers are numbers that follow in order.
For example, 1, 2, 3 or 46, 47, 48, 49.
Which 3 consecutive whole numbers have a product of 336?
Write about how you worked out the answer.

© HarperCollinsPublishers 2016

Issue II – Division   S&C Volume 3

The Maths Herald

Name: Date:

🐜 Looking for Patterns

9 and 81   3 and 72   2 and 94

For each pair of numbers:
- Divide the 2-digit number by the 1-digit number.
- Reverse the 2 digits in the 2-digit number and find the difference between this number and the 1-digit number.
- What do you notice about the 2 answers?

🐜 Looking for Patterns

Emily picked up a handful of counters. She estimated that there were between 20 and 30 counters.

To find out exactly how many counters there were, she tried to arrange the counters into groups of 2. When she did this, she found that there was one counter left over.

When she tried to arrange them into groups of 4 or 5 the same thing happened – one counter was left over each time.

She then decided to arrange the counters into groups of 3. When she did this, there were no counters left over.

How many counters did Emily pick up?

© HarperCollinsPublishers 2016

🐜 What's the Problem?

Which 2-digit number has a remainder of 1 when divided by 2, 3, 4, 5 or 6?
Write about how you discovered the number.

🐜 Let's Investigate   92 ÷ 6

Look at the division calculation on the right. The answer has been worked out using 2 different methods.

$92 ÷ 6 = (60 + 32) ÷ 6$
$ = (60 ÷ 6) + (32 ÷ 6)$
$ = 10 + 5\ r\ 2$
$ = 15\ r\ 2$

```
   ___
6 ) 92
   -60    (6 × 10)
    32
   -30    (6 × 5)
     2
```
Answer: 15 r 2

How many different methods can you use to work out the answer to this division calculation?

184 ÷ 3

factor

🐜 Sports Update

Omar is running in a 12 km long-distance race.

There are 11 flags spaced at equal distances along the course of the race, from the Start line to the Finish. When Omar passes the 6th flag, how far has he run?

Issue 11 – Division

The Arts Roundup

A television station has allocated 8 1-minute adverts during the airing of a half-hour programme that costs £960 000 to produce. What is the cost of each of the 8 adverts if all the adverts cost the same and cover the total cost of the production of the programme?

A half-hour television programme is generally 22 minutes long with 8 minutes of adverts.

The money that television stations make by televising the adverts is used to help pay for making (producing) the programme. This includes paying the salaries of the cast and crew, as well as all the other expenses there are in making a television programme.

The Puzzler

The number in the box is a 2-digit number.
The sum of the 2 digits is 16.
If you divide the number by 12 you have 1 left over.
What's the number in the box?

Looking for Patterns

There are fewer than 100 apples in this crate.

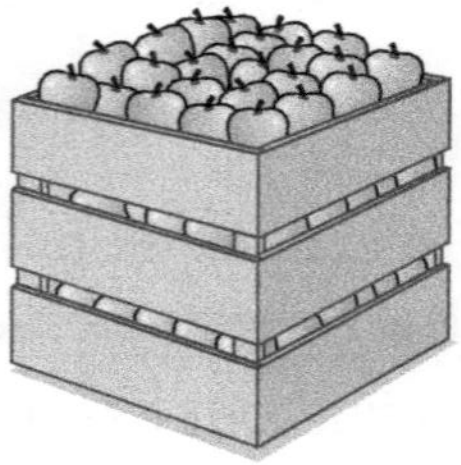

John the greengrocer can divide the crate of apples into bags of 3, 4 or 7 apples so that each bag has an equal number of apples with no apples left over.
How many apples are in the crate?

Let's Investigate

Step 1: Write down a 3-digit number in which 2 of the digits are the same.   717

Step 2: Add together all the digits in the 3-digit number.   7 + 1 + 7 = 15

Step 3: Write down all the possible 2-digit numbers you can make using each of the digits in the 3-digit number. (There should be 3 different 2-digit numbers.)   71, 17, 77

Step 4: Add together the 3 2-digit numbers.   71 + 17 + 77 = 165

Step 5: Divide the answer to Step 4 by the answer to Step 2.   165 ÷ 15 = ?

Repeat the steps above several times, starting with a different 3-digit number each time. (Make sure that 2 of the digits are the same.)
What do you notice?

What's the Problem?

A knight is in command of a troop of men and horses.
Altogether there are 60 heads and 160 legs.
How many horses does the knight have under his command?

© HarperCollinsPublishers 2016

Issue 12 – Division

S&C Volume 3

The Maths Herald

Name:

Date:

What's the Problem?

Rada takes her dog to a dog training class.

There are 76 legs and 26 heads altogether.

How many dogs and how many people are there in the class?

Focus on Science

Peter can heat his house using a gas boiler or a log stove.

He prefers to use the log stove as it does not contribute to global warming.

He buys 6 tonnes of logs in October and uses 80 kg of logs each day to heat his house.

How many days will his supply of logs last for?

The Arts Roundup

The Boehampton Amateur Dramatic Society (BADS) has between 50 and 70 members.

When all the members of BADS turn up for rehearsal, they are able to do different warm-up exercises either in pairs, or in groups of 3, 4, 5 or 6, with no one being left out.

How many members are there in BADS?

© HarperCollins *Publishers* 2016

Issue 12 – Division

At Home

Next time you visit a supermarket look out for the way in which different foods and other items are sold.

Quite often items are not sold singly (just one), but rather in multi-packs.

What are the different quantities that different items are sold in?

What is the highest quantity you can find for a multi-pack? Why do you think this is?

Let's Investigate

$165 \div 5 = ?$

Can you work out the answer to this calculation quickly in your head? Most people couldn't!

Look at this quick method for working out the answer mentally.

180 divided by 5 is......

$165 \div 5 = ?$

double

$330 \div 10 = 33$

Investigate whether this method works for dividing any multiple of 5 by 5.

Can you impress your friends by quickly telling them the answers to these calculations?

$95 \div 5 =$ $180 \div 5 =$ $425 \div 5 =$ $715 \div 5 =$

4

Issue 12 – Division

What's the Problem?

Let's Investigate

These 12 counters have been arranged to make different-shaped rectangles.

Each rectangle can be expressed as a multiplication calculation.

What calculations can you make using 10, 11, 13, 14, ... 24 counters?

1 × 12

2 × 6

4 × 3

Kylie has 4 pairs of blue socks and 5 pairs of red socks mixed up in a drawer.

What is the minimum number of socks that she needs to take out, without looking, to be certain of getting a matching pair?

Looking for Patterns

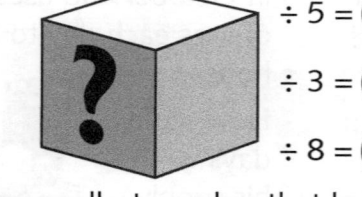

- ÷ 5 = ◯ r 4
- ÷ 3 = ◯ r 1
- ÷ 8 = ◯ r 2

What is the smallest number that has:
- 4 left over (a remainder of 4) when divided by 5
- 1 left over (a remainder of 1) when divided by 3
- 2 left over (a remainder of 2) when divided by 8?

Let's Investigate

All of the numbers below can be divided exactly by 2.

2, 4, 6, 8, 10, 12, 14, 16, ...

All of these numbers can be divided exactly by 4.

4, 8, 12, 16, 20, 24, 28, 32, ...

Continue each list.

Which numbers have a units digit of 2?
Which numbers have a units digit of 4?
What patterns do you notice?
What if you investigated numbers that can be divided by 3 and 6?
What about 4 and 8?

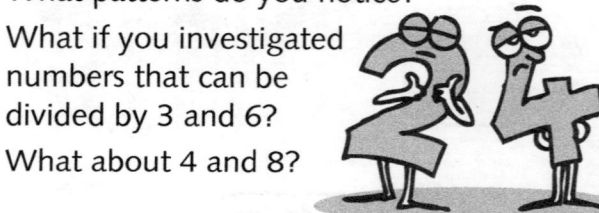

Looking for Patterns

What is the smallest number that is exactly divisible by 6, 7 and 8?

Looking for Patterns

Using a set of 0–9 digit cards it is possible to arrange some of the cards to make numbers that are multiples of 6.

For example:

| 6 | 1 | 2 | 3 | 0 | 5 | 4 | 7 | 8 |

Can you make other sets of numbers that are multiples of 6?

Investigate making multiples of other numbers.

© HarperCollins Publishers 2016

Issue 13 – Mixed operations
S&C Volume 3

The Maths Herald

Name: Date:

Money Matters

Money has been used for centuries. Before metal and paper were used to make coins and notes, people used other objects such as pebbles and shells as currencies.

Imagine that 4 shells (s) are equal to 1 pence (p).

How many shells equal:
- 2 pence?
- 5 pence?
- 10 pence?
- 20 pence?
- 50 pence?

Would the value of 360s be more than or less than 85p?

Think of 5 items. Give each item a price in pence. What is the equivalent price in shells?

What's the Problem?

What is the correct answer to each of these calculations? Explain why.

$5 + 3 \times 7 =$	$20 - 4 \times 3 =$	$12 - (5 + 6) + 4 =$	$5 \times 8 - 25 \div 5 =$
56 26	48 8	5 17	3 35

© HarperCollins*Publishers* 2016

Issue 13 – Mixed operations

The Puzzler

Write an operation and a number in each empty box to complete these calculation chains.

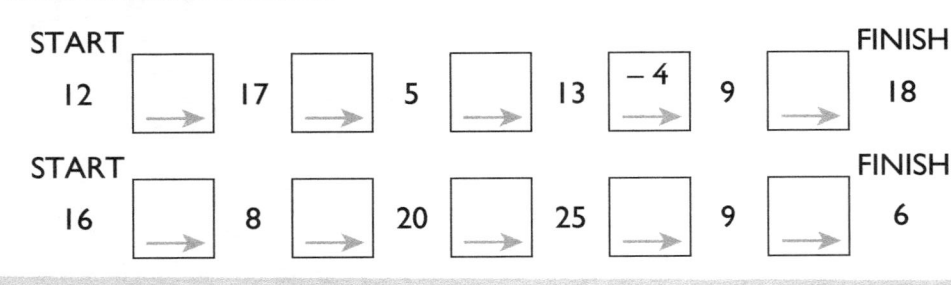

START
12 → □ → 17 → □ → 5 → □ → 13 → −4 → 9 → □ → FINISH 18

START
16 → □ → 8 → □ → 20 → □ → 25 → □ → 9 → □ → FINISH 6

What's the Problem?

A double-decker bus can seat 62 passengers.

When the bus leaves the garage there are no passengers on the bus.

At the 1st stop, 1 person gets on the bus.
At the 2nd stop, 2 people get on the bus.
At the 3rd stop, 3 people get on the bus.
This continues until the bus is full.

After how many stops will the bus be full?

Will all the people at the last bus stop where the bus becomes full be able to get a seat on the bus?

If not, how many people will not get a seat?

order of operations

Note: No one ever gets off the bus!

Issue 13 – Mixed operations

Money Matters

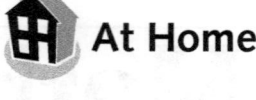

What is the cost of a 3-bedroom house and a 2-bedroom flat in your area?

What is the average house / flat price in your area?

How have prices changed over the past 12 months?

How do house prices in your area compare with those in a different area nearby?

At Home

The next time you visit a supermarket, look out for items that are on sale.

What is the difference in price between the sale price and the normal price?

How much of a saving is this?

Which items offer a "big saving"?

The Language of Maths

Look at the calculations on the right.

When a number is multiplied by itself, the answer, or product, is called a *square number*.

We can write such calculations using the abbreviation: ²

$2 \times 2 = 4$	$2^2 = 4$
$3 \times 3 = 9$	$3^2 = 9$
$4 \times 4 = 16$	$4^2 = 16$
$5 \times 5 = 25$	$5^2 = 25$
$6 \times 6 = 36$	$6^2 = 36$

Work out the answer to each of these calculations.

$7^2 =$ $10^2 =$

$19^2 =$ $12^2 =$

$30^2 =$ $80^2 =$

Money Matters

Do you think you should receive pocket money?

Realistically, how much pocket money do you think you should receive each week?

Write a list of reasons explaining why you think you should receive pocket money, and why you should receive the amount that you think is "fair".

Be prepared to justify your reasoning.

Sports Update

Malcolm is trying to get fit. Each day this week he has spent 5 minutes more on the running machine than the day before. On Sunday he spent 40 minutes on the machine.

How long did Malcolm spend on the machine on the Monday before?

The Puzzler

Write a number in each empty box to complete these calculation chains.

START → +5 → □ → −3 → □ → +10 → 20 → −6 → □ → +9 → 23 FINISH

START → □ → −7 → □ → +12 → □ → −4 → □ → +8 → □ → −16 → 9

© HarperCollins*Publishers* 2016

The Maths Herald

S&C Volume 3

Name: Date:

Technology Today

Leo, Giles and Toby each have a number of computer games.

Leo has 3 times as many computer games as Giles.

Giles has 4 less computer games than Toby. Altogether the 3 boys have a total of between 35 and 40 games.

How many computer games does each boy have?

Money Matters

Michael and Allison each have a moneybox in which they save their spare change.

Allison has saved 3 times as much money as Michael. However, they worked out that if Allison gave Michael £8, they would both have the same amount of money.

How much money has Allison saved? What about Michael?

Write about how you worked out the answer.

© HarperCollins Publishers 2016

The Puzzler

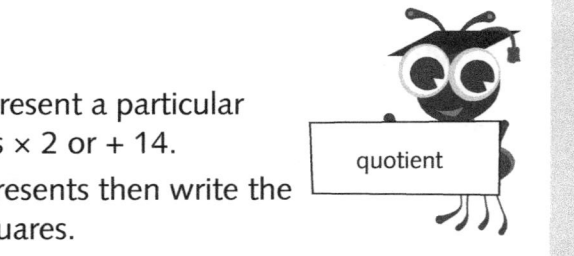

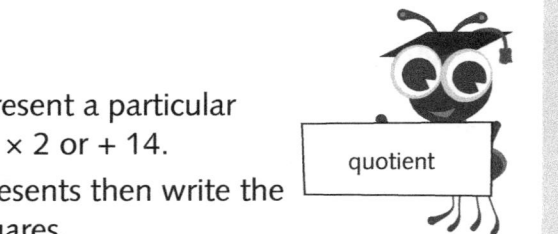

The shapes □ ◇ ○ △ each represent a particular mathematical operation such as × 2 or ÷ 14.

Work out what each shape represents then write the missing numbers in the grey squares.

quotient

4	□	16	◇	34	○	17	△	11
7	□	28	◇	46	○	23	△	17
5	□	20	◇		○	19	△	19
8	□		◇	58	○		△	9
11	□		◇		○		△	
	□		◇		○		△	
	□		◇	33	○		△	

Issue 14 – Mixed operations

Looking for Patterns

Complete the grid.

☐	◯	+ ◯	× ◯
5	9	14	45
8	7		12
7		7	70
	6	20	96
6		21	98
			90

Money Matters

Without working out the exact amounts, which of the following saving plans would give you the most money at the end of the year? Why?

- If you planned to save 20p every day for a whole year, what would be your total savings for the year?
- What if you saved £1.50 a week?

Now do the calculations to see if you chose the best plan.

What's the Problem?

I have a total of 3 brothers and sisters. Together, their ages add up to 16 years. When multiplied together they come to 140 years. How old are my brothers and sisters?

What's the Problem?

Mr and Mrs Herne have 5 children. Each son or daughter has married, and each couple have had 3 children.

2 of Mr and Mrs Herne's grandchildren are also married and 1 of these couples has a child.

How many people are there in the Herne family?

Draw a diagram to show your answer.

Money Matters

SPECIAL OFFER! Pack of 4 pencils only 60p.

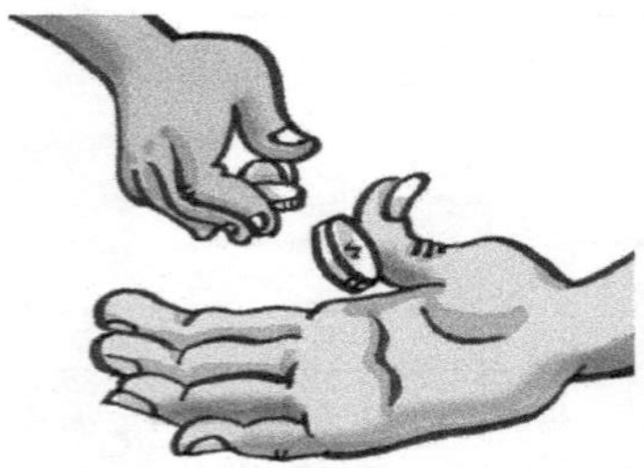

Stanley has 9 coins in his hand. With these 9 coins he has the exact amount of money needed to buy 1 pack of pencils.

What coins does Stanley have in his hand?

The Puzzler

$$AA \times A = BB \qquad AA + A = AB$$

Each of the letters stands for a different digit. Which digit does each stand for? Explain how you worked out what each letter stands for.

© HarperCollinsPublishers 2016

Issue 15 – Mixed operations

S&C Volume 3

The Maths Herald

Name:

Date:

The Puzzler

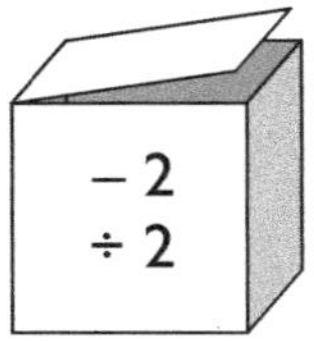

+ 2
× 2

Which number gives the same result if you add 2 to it or multiply it by 2?

− 2
÷ 2

Which number gives the same result if you take 2 from it or divide it by 2?

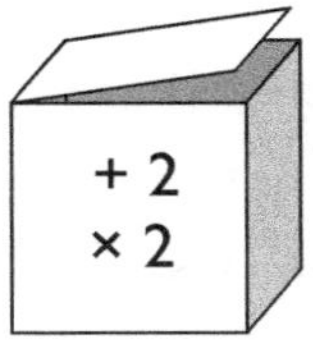

Sports Update

Surinder was asked how many points she had scored playing basketball this season. She said:

"I've scored 3 times as many points as Alex, and Tom has scored twice as many points as me, less 5 points. Between us, we have scored 135 points."

How many points has Surinder scored?

Technology Today

Memory size in a computer is measured in *bytes*.

One byte is the unit that stores one character, such as a letter of the alphabet, a 1-digit number, a mathematical symbol or a punctuation mark.

Graphics such as illustrations and diagrams use a larger number of bytes.

This booklet was written on a computer. Excluding the graphics, approximately how many bytes were needed to store all the characters on these 4 pages?

© HarperCollins*Publishers* 2016

What's the Problem?

Altogether there are 80 people attending the wedding reception of Oscar and Lucinda.

Using a total of 11 tables only and having no spare seats, Wondrous Weddings can seat everyone on small tables for 6 people and large tables for 8 people.

How many of each size table do they use?

Sports Update

In the game of tenpin bowling, the pins are arranged in 4 rows as shown in the picture below.

Imagine a giant game of bowling with 6 rows of pins.

How many pins would be used in the game?

What about in a monster game of bowling that has 9 rows of pins?

Looking for Patterns

24 children are standing in a circle playing a game.

One child starts counting from 1 round the circle. As the count goes round the circle, every 2nd child sits down.

How many children are still standing after 2 counts round the circle?

How many children are still standing after 3 counts round the circle?

Issue 15 – Mixed operations

Let's Investigate

- Choose a whole number between 1 and 50.
- Subtract this number from 100.
- Divide the answer by 10.
- Subtract this number from 10.
- Multiply the answer by 100.
- Subtract this number from 1000.
- Divide the answer by 10.
- Subtract this number from 100.

What do you notice?
Why does this happen?

Let's Investigate

Use the digits 2, 5 and 8 to write as many different 3-digit numbers as possible. Investigate all the different sums and differences possible for pairs of numbers. Copy this table and write each calculation in the appropriate section.

Answer between	Calculation
1–200	
201–400	
401–600	
601–800	
801–1000	
1001–1200	
1201–1400	
1401–1600	
1601–1800	

Focus on Science

At birth, the human body consists of about 270 bones. But, as we grow older some of the bones fuse together.

The adult human skeleton consists of 206 bones that vary in size and shape. The diagram on the right shows how these bones are distributed throughout the human body.

All these bones together make the skeleton and support our body and maintain its shape.

Look carefully at the skeleton and write statements comparing the number of bones in different parts of the body. For example:

- Our skull has 7 times as many bones as our shoulder girdle.
- Over half the bones in our body are in our hands and feet.

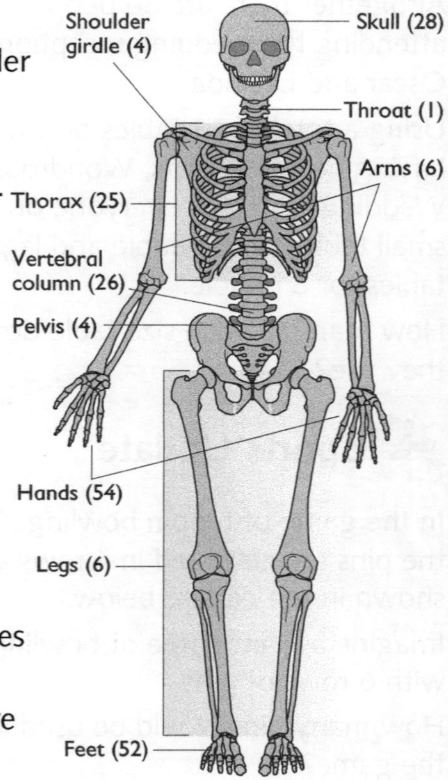

Skull (28)
Throat (1)
Shoulder girdle (4)
Thorax (25)
Vertebral column (26)
Arms (6)
Pelvis (4)
Hands (54)
Legs (6)
Feet (52)

What's the Problem?

Lakshmi has 2 brothers and a sister. Her sister is older than both her brothers. The product of the ages of the younger 2 equals the age of the older one. The sum of all 3 ages is equal to the product of the ages of the youngest and the oldest. How old are her brothers and sister?

Issue 16 – Mixed operations

S&C Volume 3

The Maths Herald

Name: Date:

What's the Problem?

Drisse bought 5 chocolate bars.

He spent a total of £3.35.

What did Drisse buy?

CHOC-BLOC 60p

Choco-Melto 45p

Chocolate Delights 85p

Let's Investigate

Write as many different calculations as you can that have an answer of 12.

Be as creative as you can.

Now sort all your different calculations.

How else could you sort your calculations?

and the answer is....

The Puzzler

Imagine rolling a 1 to 6 dice and adding together all the numbers rolled.

You are allowed to either use the number rolled, or multiply the number rolled by 10 and use this number instead.

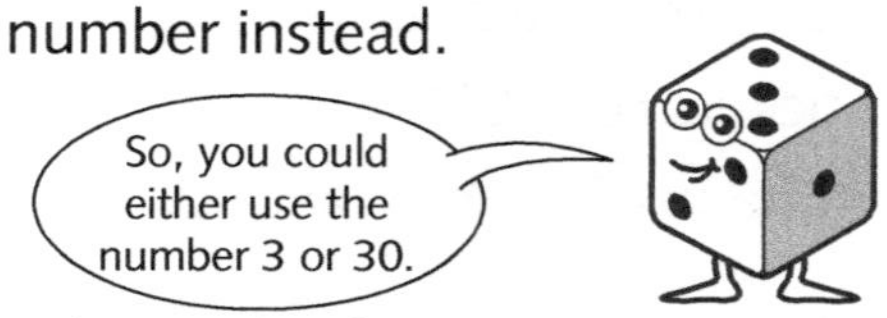

So, you could either use the number 3 or 30.

What is the fewest number of rolls you would need to make a total of 99?

The Puzzler

$A + B + C = D$
$A \times B \times C = D$

Each of the letters stands for a different digit. Which digit does each stand for? Explain how you worked out what each letter stands for.

The Arts Roundup

Newspapers, particularly weekend newspapers, include a "What's On" guide, listing all the different types of entertainment that is currently on or that is about to start.

The list on the right includes some of the different types of entertainment.

Investigate different types of entertainment.

Which is generally the cheapest type of entertainment? Which is the most expensive? Why do you think this is?

- Movies
- Theatre
- Musicals
- Dance
- Concerts
- Opera

Let's Investigate

1, 2, 3, 4

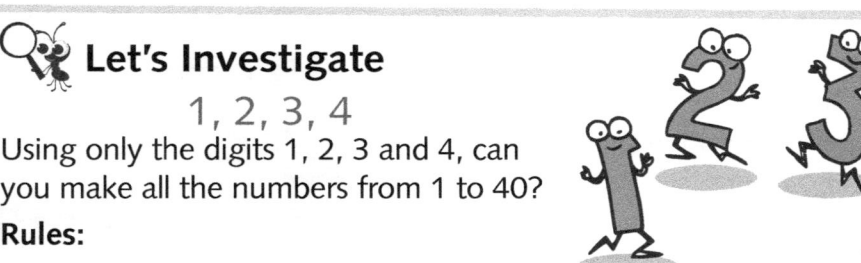

Using only the digits 1, 2, 3 and 4, can you make all the numbers from 1 to 40?

Rules:

- Each digit can only be used once in each calculation.
- You can arrange 2 of the digits into a 2-digit number.

Here are 2 to get you started:

$10 = (2 \times 3) + 4$

$32 = 34 - 2$

© HarperCollins Publishers 2016

Issue 16 – Mixed operations

Let's Investigate

A postal worker has 6 letters to deliver between 6 houses. All 6 letters might be for 1 house, or there might be 1 letter for each house, or any other combination.
How many different ways can there be for the postal worker to deliver all 6 letters to the 6 different houses?

Money Matters

Sara is taking some children and adults to the cinema.
The tickets cost £71 in total.
How many of each type of ticket does she buy?
Write about how you worked out the answer.

Adults £12
Children £7

The Puzzler

When can you add 2 to 11 and get 1?

? ? ? ?
11 + 2 = 1

The Language of Maths

Henry is looking at photo of a man. He says: "Brothers and sisters have I none, but this man's father is my father's son."
Who is Henry looking at?

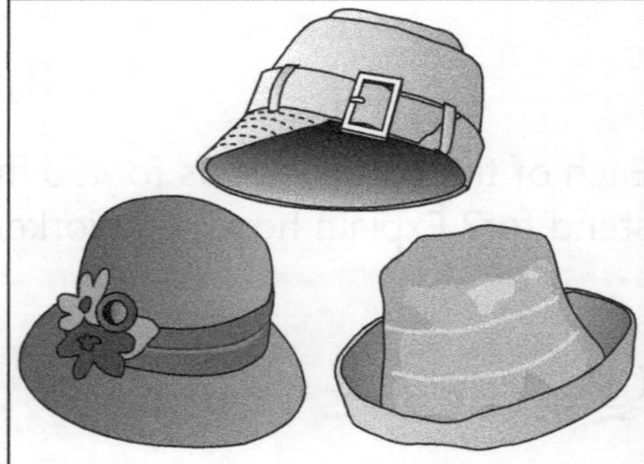

2 mothers and 2 daughters went shopping.
They each bought a hat, but they only bought 3 hats in total.
How come?

What's the Problem?

Ben can build a boat in 12 days by himself.
Noah can build a boat in 6 days by himself.
How long would it take them to build a boat together?

The Puzzler

Which 2 numbers have a difference of 9 and a sum of 23?

difference

© HarperCollins*Publishers* 2016

Issue 17 – Fractions
S&C Volume 3

The Maths Herald

Name: _____ Date: _____

Let's Investigate

Imagine that you have shrunk to $\frac{1}{10}$ of your normal height. What would the measurements be for each of these pieces of clothing?

If you're a boy
- a pair of trousers
- a shirt

If you're a girl
- a skirt
- a shirt

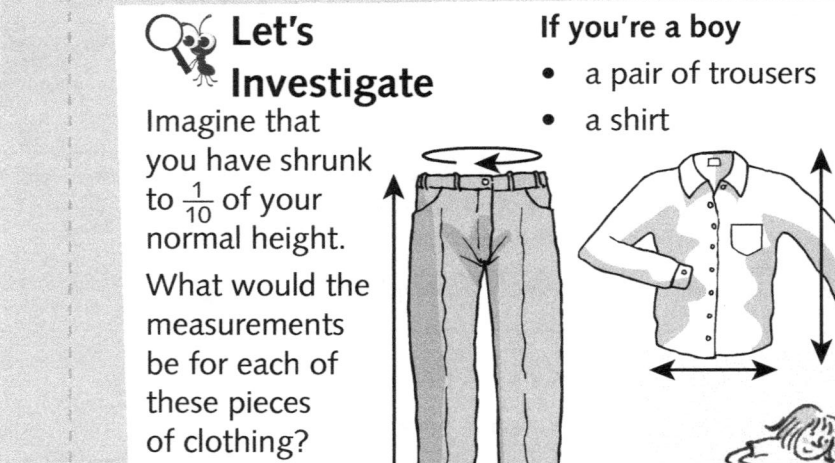

The Puzzler

A king had 4 daughters. On his deathbed he said:

> To my eldest daughter, I leave one-quarter of my kingdom.
> To my 2nd daughter, I leave two-eighths of my kingdom.
> To my 3rd daughter, I leave three-eighths of my kingdom.
> And to my youngest daughter, I leave the rest of my kingdom.

How much did his youngest daughter inherit?

© HarperCollins*Publishers* 2016

Construct

Using interlocking cubes, construct a model that has:
- half of its cubes red
- one-third of its cubes blue
- one-eighth of its cubes green
- the rest of its cubes yellow.

Can you construct more than one model using these fractions?

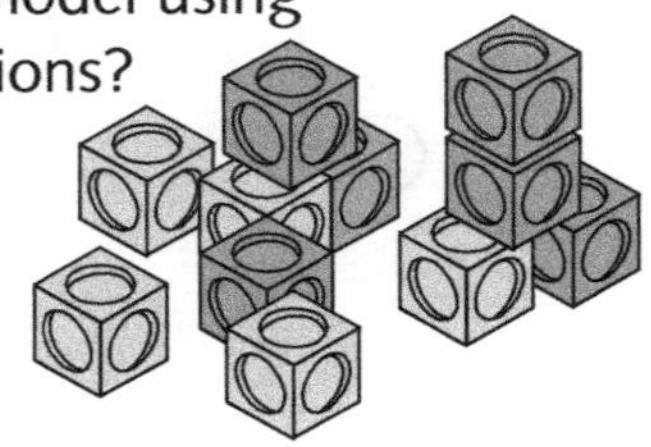

What's the Problem?

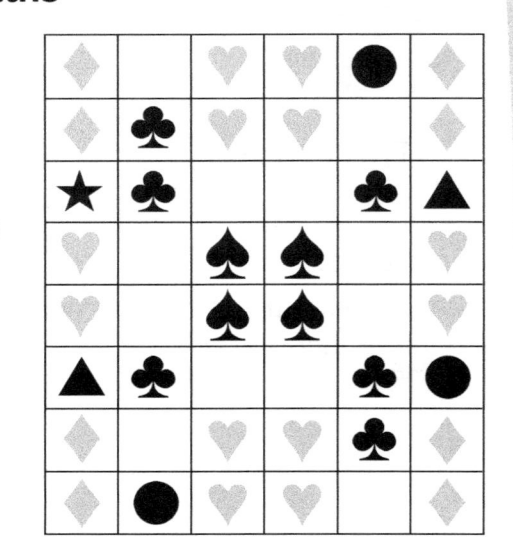

Jamila sleeps for 9 hours each night. Her friend Monica sleeps for 10 hours each night.

Jamila says to Monica: "You sleep 15 days more than I do each year."

Is she correct?

Write about how you worked out the answer.

The Language of Maths

Write different fraction statements comparing the different shapes in this grid. Here are 2 to get you started.

- One-quarter of the squares on the grid have a heart in them.
- Half as many squares on the grid have a circle in them when compared to squares with clubs in them.

Issue 17 – Fractions

Let's Investigate

Here is one way to share 2 chocolate bars equally between 3 people.

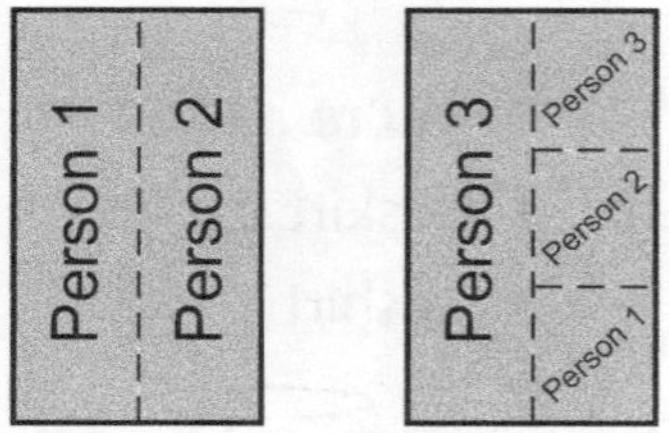

| Person 1 |
| Person 2 |

Here is one way to share 3 chocolate bars equally between 4 people.

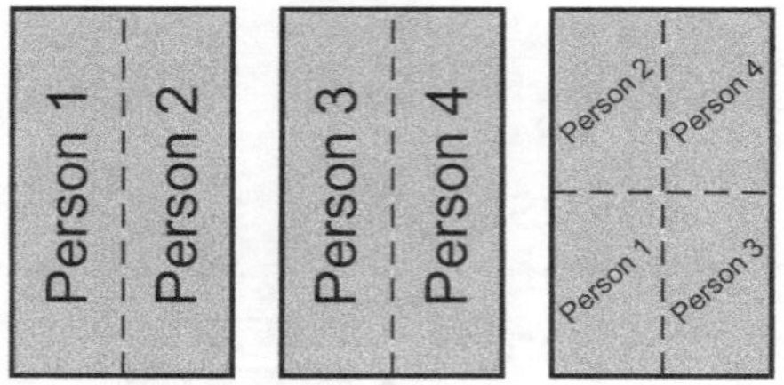

| Person 1 |
| Person 2 |
| Person 3 |
| Person 4 |

Show ways to share equally the following chocolate bars between these people:

- 2 chocolate bars between 5 people
- 3 chocolate bars between 5 people
- 3 chocolate bars between 6 people
- 3 chocolate bars between 8 people
- 3 chocolate bars between 10 people
- 5 chocolate bars between 8 people

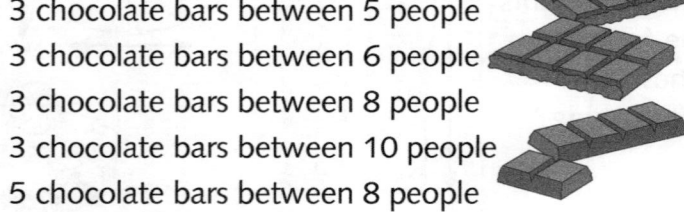

What's the Problem?

Annabelle invited some friends to a party. This photo shows two-thirds of the friends she invited. How many friends did Annabelle invite to her party?

The Puzzler

Colour $\frac{1}{6}$ of the circles blue.
How many blue circles is this?
Now colour $\frac{1}{4}$ of the total number of circles red.
How many red circles is this?
How many circles are not coloured?
What fraction of the circles is coloured? What fraction of the circles is not coloured?
Now colour circles green so that only $\frac{1}{12}$ of the circles are still not coloured.
How many circles are coloured now?
Now what fraction of the circles is coloured?

numerator and denominator

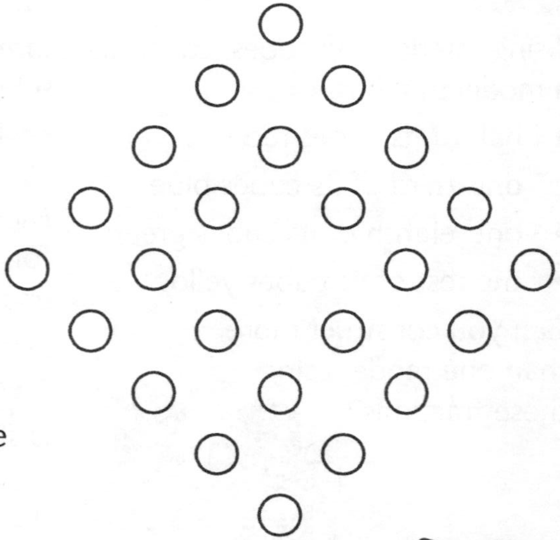

The Puzzler

One-quarter of this square has been shaded.
Can you divide the remaining white part into 4 identical pieces?

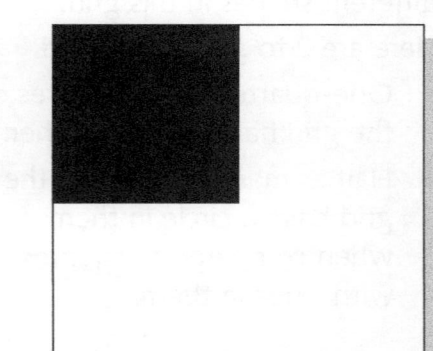

© HarperCollins*Publishers* 2016

Issue 18 – Fractions
S&C Volume 3

The Maths Herald

Name:

Date:

🐛 The Language of Maths

Fold a sheet of A4 paper in half.

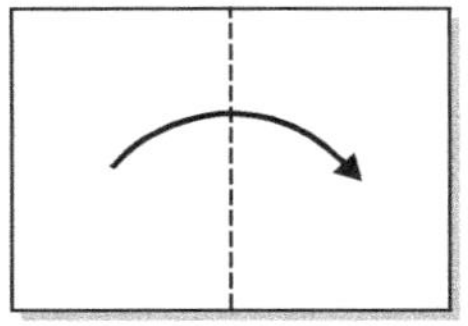

Fold it in half again.

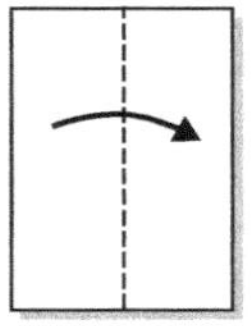

Fold it in half once more.

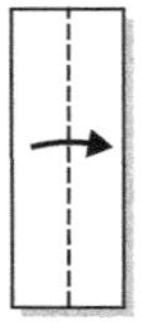

Open out the sheet of paper and label the sections like this:

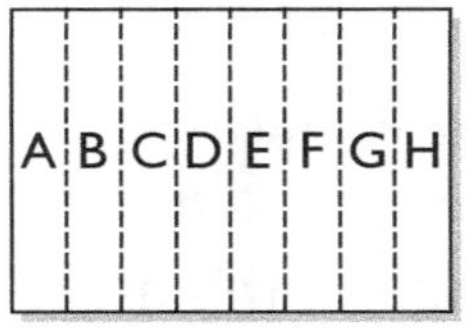

| A | B | C | D | E | F | G | H |

Write statements about the fractions of different sections and combinations of sections.

For example:
- Section A is $\frac{1}{8}$ of the sheet of paper.
- Sections A and B combined are $\frac{1}{4}$ of the sheet of paper.

🧩 The Puzzler

$\frac{1}{2}$ said to $\frac{1}{5}$: "Go and find another fraction to make us 1."

What fraction did $\frac{1}{5}$ find?

© HarperCollins*Publishers* 2016

🐛 The Language of Maths

The diagram on the right is called a *fraction wall*. It is used to compare fractions and identify fractions that are equivalent (the same).

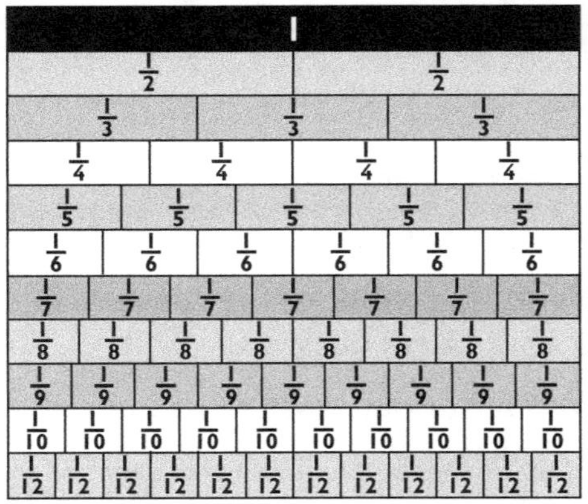

For example:
- two-thirds is less than three-quarters
- two-fifths is equivalent to four-tenths.

Using the terms *more than*, *less than* and *equivalent*, write statements comparing fractions and finding equivalent pairs of fractions.

unit fraction

🐛 The Language of Maths

Look around your classroom.

What statements that involve fractions can you write to describe what you see around you?

You can use words such as *about, almost, approximately, just over, just under, slightly more than, slightly less than*.

Here are 3 to get you started:
- Half the children that sit at my table are girls.
- Almost two-thirds of the children in our class have school dinners.
- Seven-tenths of our windows are open.

Issue 18 – Fractions

Let's Investigate

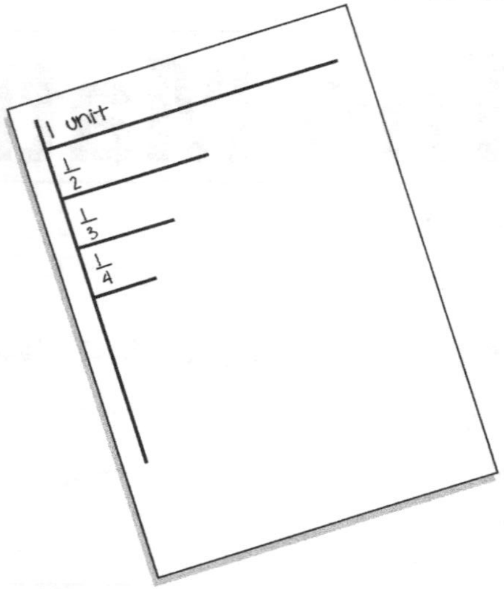

At the far left-hand side of a sheet of paper, using a ruler, draw a vertical line down the length of the sheet.

Now draw a horizontal line 20 cm long and label the line "1 unit".

Below this line draw another line that is half the length and label it "$\frac{1}{2}$".

Draw a third line that is one-third the length and label it "$\frac{1}{3}$".

Then draw a fourth line that is one-quarter the length and label it "$\frac{1}{4}$".

Keep going like this as far as you can.

When you have gone as far as you can, join the right-hand end of all the lines to make a curve.

Notes:
- Make sure that all the horizontal lines start from the vertical line.
- Make sure that all the lines are the same distance apart.

Let's Investigate

$\frac{1}{2}$ means one divided by two (1 ÷ 2).

What happens when you key this division calculation into a calculator? What about $\frac{1}{3}$?

Investigate what happens when you key in all the unitary fractions to $\frac{1}{12}$.

What is the name we give to these numbers?

Write down the answers and write about any patterns you notice in the answers.

Let's Investigate

$$\frac{1}{10}, \frac{2}{10}, \frac{3}{10}, \frac{4}{10}, \frac{5}{10}, \frac{6}{10}, \frac{7}{10}, \frac{8}{10}, \frac{9}{10}$$

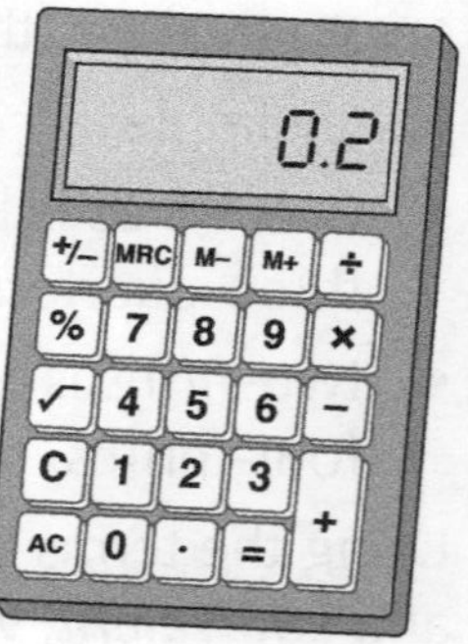

These fractions are all called *tenths*. Can you see why? Remember that $\frac{1}{10}$ means one divided by ten (1 ÷ 10).

What happens when you key this division calculation into a calculator?

What happens when you key $\frac{2}{10}$ (2 ÷ 10) into a calculator?

Without keying in any more calculations, predict what would happen if you keyed in all the tenths to $\frac{9}{10}$.

Let's Investigate

$$\frac{1}{100}, \frac{2}{100}, \frac{3}{100}, \ldots \frac{17}{100}, \ldots \frac{32}{100}, \frac{65}{100}, \ldots \frac{84}{100}, \ldots \frac{99}{100}$$

These fractions are all called *hundredths*. Can you see why?

Remember that $\frac{1}{100}$ means one divided by one hundred (1 ÷ 100).

What happens when you key this division calculation into a calculator?

What happens when you key $\frac{2}{100}$ (2 ÷ 100) into a calculator?

Without keying in any more calculations, predict what would happen if you keyed in all the hundredths to $\frac{99}{100}$.

Issue 19 – Fractions
S&C Volume 3

The Maths Herald

Name:
Date:

Focus on Science

The size of diamonds, and other precious gemstones like rubies and emeralds, is measured in *carats* (ct).

This actually measures the weight of the diamond, one carat being $\frac{1}{5}$ g in weight.

Therefore a 5-carat diamond is a diamond that weighs $5 \times \frac{1}{5}$ g = 1 g.

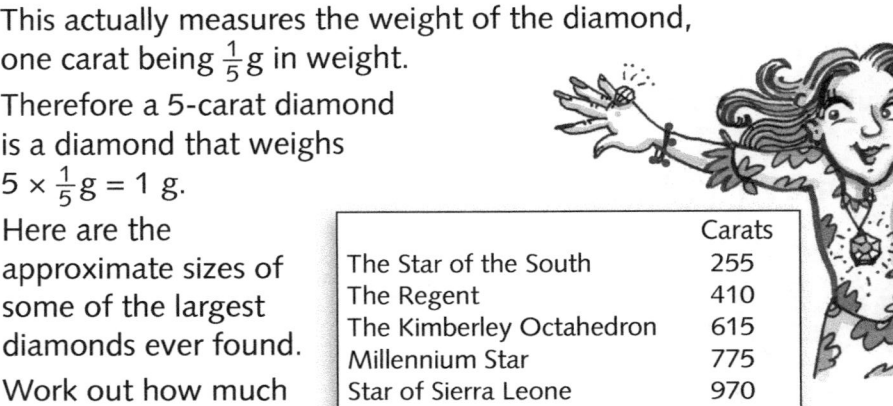

Here are the approximate sizes of some of the largest diamonds ever found. Work out how much each of them weighed.

	Carats
The Star of the South	255
The Regent	410
The Kimberley Octahedron	615
Millennium Star	775
Star of Sierra Leone	970
Cullinan Diamond	3105

Money Matters

Alpesh has 24 coins in his money box.
Half of the coins are 50p coins.
One-third of the coins are £1 coins.
One-sixth of the coins are £2 coins.
How much money does Alpesh have in his money box?

What's the Problem?

Josh and Ellie eat a bag of 32 sweets between them.

If Ellie had eaten twice as many sweets as she did, Josh would have eaten only $\frac{2}{3}$ as much as he did.

How many sweets did each of them actually eat?

What's the Problem?

Tim feeds his dog Towser $\frac{2}{3}$ of a tin of dog food twice a day.
He buys the dog food in boxes of 12 tins.
How long does a box of dog food last Towser?

The Puzzler

When you add 2 to both the numerator and the denominator of a particular fraction it doubles in value.

What is the fraction?

$$\frac{\text{numerator}}{\text{denominator}}$$

non-unit fraction

© HarperCollins*Publishers* 2016

Issue 19 – Fractions

Money Matters

Imagine you have a job in an office and earn £12 each hour.
How much would you earn if you worked 40 hours in a week?
How much would you earn in a year if you worked for 50 weeks?

In 1 year you spend:
- $\frac{1}{2}$ on accommodation
- $\frac{3}{20}$ on food
- $\frac{3}{10}$ on holidays
- $\frac{1}{20}$ on entertainment

… and you save $\frac{1}{5}$.

Work out how much you spend on each of the above things in a year.

Money Matters

Choose 2 coins and write down the total.
What fractions of the total can you make where the amount?

For example:

Total = 12 p

$\frac{1}{2} \times 12p = 6p$ $\frac{3}{4} \times 12p = 9p$ $\frac{5}{12} \times 12p = 5p$
$\frac{1}{3} \times 12p = 4p$ $\frac{1}{6} \times 12p = 2p$ $\frac{7}{12} \times 12p = 7p$
$\frac{2}{3} \times 12p = 8p$ $\frac{5}{6} \times 12p = 10p$
$\frac{1}{4} \times 12p = 3p$ $\frac{1}{12} \times 12p = 1p$ $\frac{11}{12} \times 12p = 11p$

Choose different pairs of coins and investigate what fractions of the total you can make where the result is a whole amount.

Money Matters

Which amounts of money would you rather be given?

Would you rather … ?	Would you rather … ?	Would you rather … ?
$\frac{1}{4} \times £18$ OR $\frac{1}{3} \times £12$	$\frac{1}{5} \times £29$ OR $\frac{1}{8} \times £46$	$\frac{3}{4} \times £36$ OR $\frac{2}{3} \times £42$

		$\frac{3}{8} \times £28$ OR $\frac{5}{6} \times £15$

Make up some "Would you rather … ?" problems for a friend to solve.
Try and make them as tricky as you can!

What's the Problem?

Conroy bought a packet of biscuits and ate one-third.
His brother Daniel came along and ate half of the biscuits that were left.
If the new packet held 24 biscuits, how many biscuits are now left in the packet?

What's the Problem?

In a survey of 24 children, $\frac{3}{4}$ said that they liked lemonade, $\frac{2}{3}$ said they liked orange juice, and $\frac{1}{2}$ that they liked both lemonade and orange juice.

How many children liked neither lemonade nor orange juice?

© HarperCollins*Publishers* 2016

Issue 20 – Fractions

S&C Volume 3

The Maths Herald

Name:

Date:

What's the Problem?

Tim feeds his dog $1\frac{1}{4}$ tins of dog food each day.

He buys the tins in boxes of 12.

Each box of 12 tins cost £14.40.

How much does Tim spend on dog food in a year?

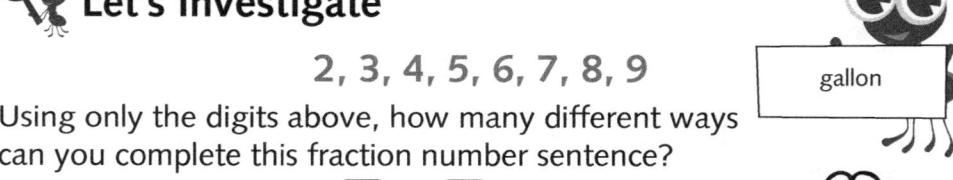

gallon

Let's Investigate

2, 3, 4, 5, 6, 7, 8, 9

$$\frac{\square}{\square} - \frac{\square}{\square} = 0$$

Using only the digits above, how many different ways can you complete this fraction number sentence?

Rules:

- You can only use the same digit once in each number sentence.
- The top digit (the numerator) must be smaller than the bottom digit (the denominator).

$$\frac{\text{numerator}}{\text{denominator}}$$

$\frac{2}{3}$ ✓ $\frac{3}{2}$ ✗

© HarperCollins*Publishers* 2016

Issue 20 – Fractions

Construct

On squared paper, draw a rectangle 12 squares by 20 squares.

Now, design a flag for your class using white and 4 other colours.

You can have whatever patterns or pictures you want on your flag, but the fractions of each colour must be as follows:

$\frac{1}{3}$ of your flag must be white, $\frac{1}{4}$ blue, $\frac{1}{5}$ red, $\frac{1}{12}$ yellow and the rest green.

What's the Problem?

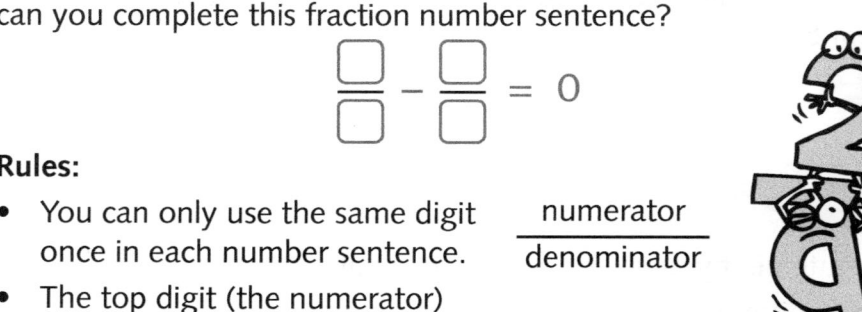

Paul fills an 8-litre watering can $\frac{3}{4}$ full. He divides the water equally between 3 plant pots.

How many litres of water does he put on each pot?

What's the Problem?

Leroy is $\frac{4}{5}$ the height of his sister Lorray, who is 27 cm taller than him.

How tall are Lorray and Leroy?

Issue 20 – Fractions

What's the Problem?

Tamsin's bedroom measures 3 m by 4 m.
$\frac{1}{3}$ of the floor space is taken up by her bed, and $\frac{1}{12}$ of the floor space is taken up by her wardrobe.
What fraction of the floor space is left free?

Let's Investigate

| 2 | 3 | 4 | 5 | 6 | 7 | 8 | 9 |

$$\frac{1}{\square} \times \frac{\square}{\square} = \frac{\square}{\square}$$

Choose 4 of the digit cards above to complete the fraction statement below.

In each statement you can only use each digit card once.
How many different fraction statements can you make?

Around the World

The Niagara Falls straddles the border between the US and Canada. 750 000 gallons of water fall over the edge every second.
$\frac{4}{5}$ of the water goes over the Canadian side of the Falls and $\frac{1}{5}$ goes over the American side of the Falls.

How many gallons of water go over the Canadian side of the Falls each second? How many gallons of water go over the American side of the Falls each second?

In the Past

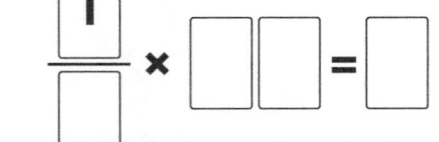

The Romans divided their armies into *legions*.
Each legion consisted of approximately 5000 men.
When the Romans occupied Britain, they stationed 3 legions in it.
Each legion was divided into 10 *cohorts*; each cohort was divided into 6 *centuria*; each centuria consisted of 80 soldiers commanded by one *centurion*.
Exactly how many fighting men were there in a legion?
What fraction of them were centurions?

© HarperCollins*Publishers* 2016

Issue 21 – Length
S&C Volume 3

The Maths Herald

Name: Date:

The Puzzler

There is a square island, $2\frac{3}{4}$ m wide, in the middle of a square pond that is $5\frac{1}{4}$ m wide. You have two planks, each $1\frac{1}{4}$ m long. Can you cross to the island without getting wet? If so, how?

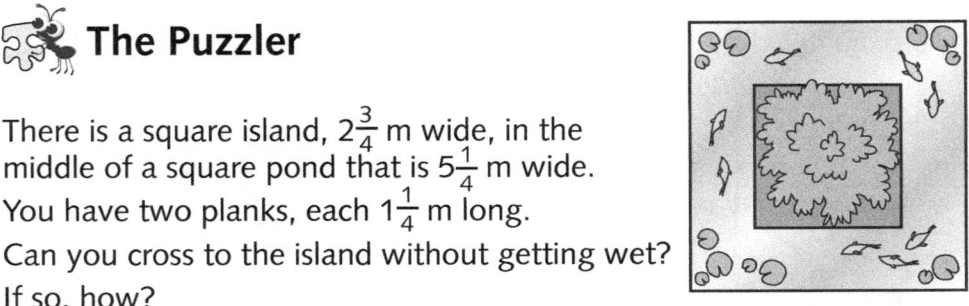

The Arts Roundup

Artists sometimes use their thumb to draw an object to scale. By holding a thumb up to an object, and using it as a type of measuring device, it makes it easier to draw the object in the correct *proportions*.

Find somewhere around the school that offers a good view of a range of things that are different heights, widths and lengths. Using a sheet of art paper and drawing pencils, draw a simple picture of what you see.

Your ability as an artist is not important for this activity. However, what is important is how accurately you are able to draw things to scale.

So don't forget to use your thumb!

© HarperCollins*Publishers* 2016

Issue 21 – Length

Around the World

The distance around a fence is 34 m.

The length of the fence is 7 m greater than the width.

What are the length and width of the fence?

Sports Update

Sam and Mats try to keep fit. Sam likes to run, while Mats prefers to cycle.

In the same amount of time, Sam can run 3 km and Mats can ride 7 km.

At this rate, how far will Sam have run if Mats has ridden 35 km?

Let's Investigate

10, 11, 12, …

Estimate how many paces it would take you to walk 10 m. Write down your estimate. Now find out exactly how many paces it takes you to walk 10 m. Write about what you did to find the actual number of paces.

Issue 21 – Length

Around the World

Look at the labels on different items of clothing.
Where was each item of clothing made?
Approximately how far did each piece of clothing travel to get to you?

What's the Problem?

Tom has climbed one-third of the way up a tree. He has another 12 metres to climb. How tall is the tree?

What's the Problem?

If everyone in your class lay down head to foot, approximately how far would you stretch?
What about everyone in the school?
Write about how you arrived at your approximations.

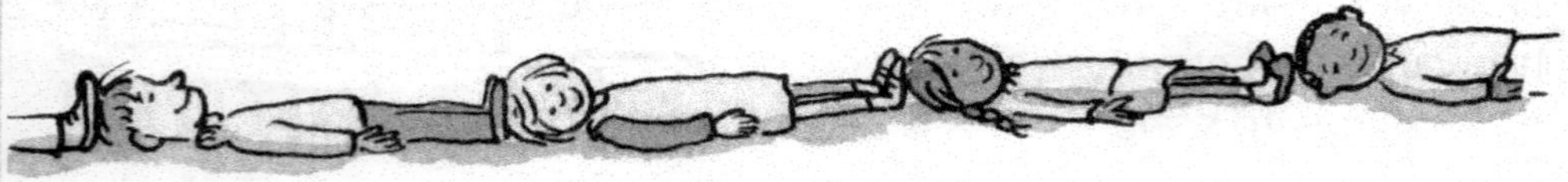

2

Issue 21 – Length

Focus on Science

How far can you and your friends see?
Using a computer, design an eyesight testing board made of lines of letters that begin large and gradually get smaller and smaller as they go lower down the page.
Decide on a distance to place the board from the friends you will be testing so that they will be able to read most of the lines on your board.
Put up your eyesight board and mark your distance line from the board.
Carry out your test.
Tell your friends to stand on the distance line and move closer to or further from the board until they are able to read each line on your board. Then measure the distance they are standing from the board.
Then write about the results of your experiment.

At Home

Road speed signs show how fast you are allowed to travel.
Investigate examples of speed signs in your local area.
Why do speed limits differ in different areas?

Let's Investigate

Approximately how many metres of continuous writing does a normal lead pencil last for?
Write about how you would find out the answer to this question.

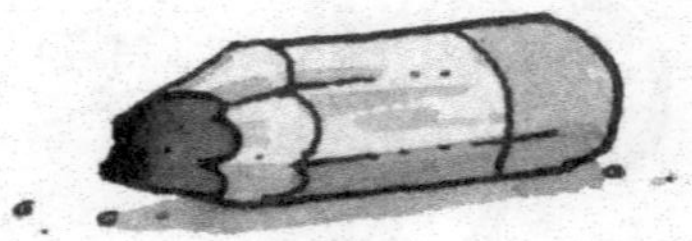

© HarperCollins*Publishers* 2016

3

Issue 22 – Mass
S&C Volume 3

The Maths Herald

Name: _____ Date: _____

Money Matters

Mollymook Resort offers freshly squeezed orange juice at breakfast.

Each week the resort uses 42 kg of oranges.

Oranges are sold in 6 kg bags at £6.40 each, or 10 kg bags at £7.50 each.

How many bags of each size must the resort buy to get the best value for money? How much will this cost them?

Construct

Hold a 500 g weight. Take time to feel how heavy the weight is.

Using modelling clay, make your own 500 g weight. Keep feeling the actual weight to make sure your weight is as close to 500 g as possible.

When you are confident that you have made a weight of 500 g, check its weight using scales.

There is very little chance that your weight will be exactly 500 g. Write down the difference between your modelling clay weight and an actual 500 g weight.

Now repeat the activity, this time holding a 100 g weight, and then making your own 100 g modelling clay weight.

Let's Investigate

How much does a handful weigh? Try this activity with a friend or in a small group.

Take a handful of something – it might be counters, interlocking cubes, beads, crayons, pencils, dice, …

Estimate how much you think the handful will weigh.

Ask your partner or the other people in your group to take a handful of the same object.

Now each of you weigh your handful.

Repeat for handfuls of different objects.

Whose estimate was nearest to the actual weight?

Did you make different estimates and get different readings? Why do you think this is?

What's the Problem?

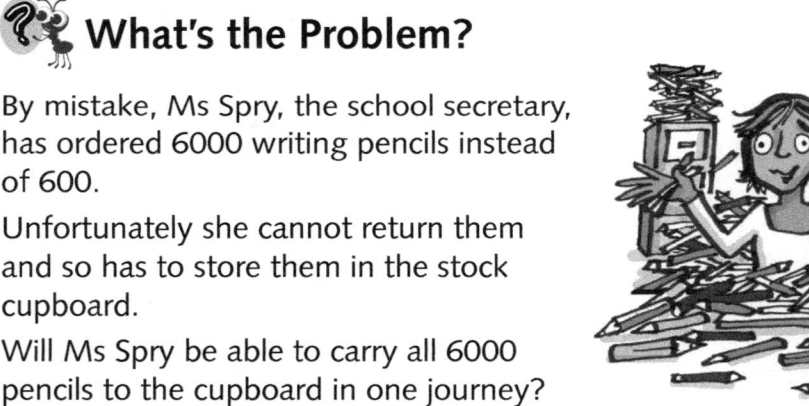

By mistake, Ms Spry, the school secretary, has ordered 6000 writing pencils instead of 600.

Unfortunately she cannot return them and so has to store them in the stock cupboard.

Will Ms Spry be able to carry all 6000 pencils to the cupboard in one journey?

© HarperCollins*Publishers* 2016

Issue 22 – Mass

Focus on Science

Which of your fingers is the strongest?
What is the maximum mass you can lift with one finger?
Design an experiment to find out.
Display your results in a table.
Were your predictions correct?
Now write about your experiment.

What's the Problem?

A customer goes into a greengrocer and asks for 1 kg of beans.
The grocer says:

> I'm sorry, my weighing scales are not working properly. One side weighs more than the other.

approximation

The customer asks if the grocer has a 1 kg weight and some lead beads.
The grocer produces them and they weigh 1 kg of beans.
How do they do it?

Looking for Patterns

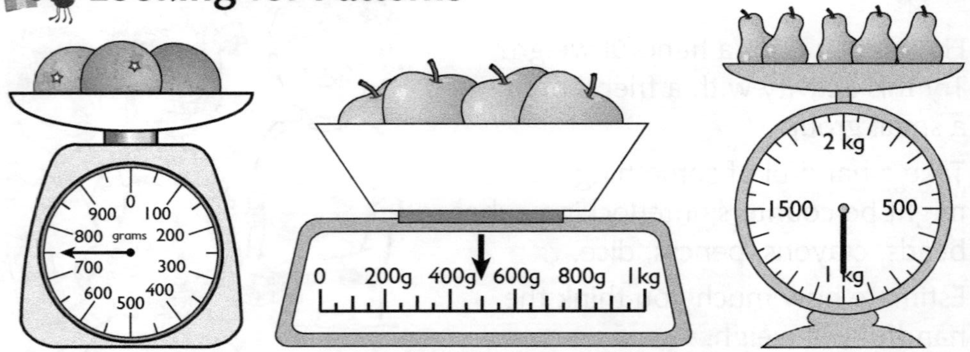

What is the total mass of 1 orange, 1 apple and 1 pear?
What is the total mass of 2 oranges, 1 apple and 2 pears?
How many oranges have the same mass as 6 apples?
Joan wants to buy exactly $2\frac{1}{2}$ kg of fruit. She wants to buy more or less the same number of each fruit. How can she do this?

What's the Problem?

Flour is sold in 3 different sizes.
One sack of flour has a mass the same as 5 large bags of flour or 20 small bags of flour.
How many small bags of flour have a mass the same as 1 large bag of flour?

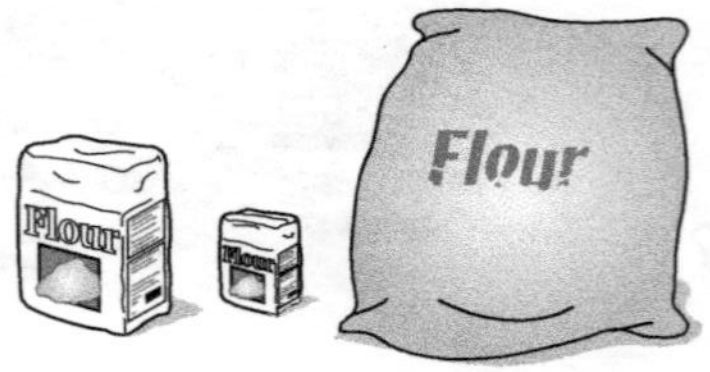

What's the Problem?

Tim's dog Towser has a mass of 9 kg plus two-thirds of its own mass. What is Towser's mass?

© HarperCollinsPublishers 2016

Issue 23 – Capacity and volume

S&C Volume 3

The Maths Herald

Name:

Date:

At Home

In supermarkets, liquids are often packaged in non-standard amounts, that is, not in litres, $\frac{1}{2}$ litre or $\frac{1}{4}$ litre, but in 330 ml or 660 ml.

In what amounts are different types of liquids packaged?

Are milk, fruit juices and soft drinks packaged in the same units?

What about washing liquids and detergents?

Are food items packaged in different amounts to cleaning items?

Investigate, and suggest possible reasons for your findings.

The Puzzler

6 glasses are in a row. The first 3 are full of water and the others are empty.

By moving only 1 glass, how can you have alternating full and empty glasses?

justify

© HarperCollins*Publishers* 2016

Focus on Science

An activity to do with a friend.

Using 2 sheets of A4 paper make 2 different cylinders:
- a tall thin cylinder by rolling the paper lengthways
- a short fat cylinder by rolling the paper widthways.

Which cylinder do you think holds the most? Why?

Now find out by filling both cylinders to the top using the same objects – perhaps small counters or cubes.

Count the number of objects in each cylinder.

What did you find out?

Was your prediction correct?

Hints:
- When making the cylinders, tape the edges together without overlapping the paper.
- One person needs to hold the cylinder steady while the other person carefully fills the cylinder.

Let's Investigate

Imagine you are going to give a party. There will be 12 people at your party – this includes yourself.

Each person will want to have about 3 drinks.

How many bottles of drink do you think you will need to buy?

Be prepared to justify your decisions with evidence.

Issue 23 – Capacity and volume

What's the Problem?

A woman came to a farmer and wanted to buy 2 litres of milk. The farmer only had a 3-litre jug and a 4-litre jug.

How could the farmer measure out the milk for the woman?

The Puzzler

Imagine a jug of water. If you double all the jug's measurements, how much water would it hold?

Focus on Science

Using a dishwasher saves water compared to washing up by hand.

Investigate the statement above. Be prepared to justify your answer.

Let's Investigate

Measure a litre of water.
Estimate what its mass will be.
Now check its mass.
How accurate was your estimate?
Repeat, measuring a litre of something else.
Do you think it will have a larger or smaller than a litre of water? Why?

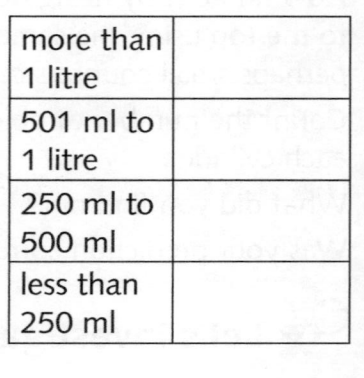

At Home

Do not open any of the containers in this activity and wash your hands afterwards.

Look around your home for different containers that hold liquid.

Good places to look include the bathroom and kitchen. If you have a laundry room, a garage or a shed, then these are also good places to look.

Look at the units that the contents of the containers are measured in.

Draw up a list similar to the one on the right.

more than 1 litre	
501 ml to 1 litre	
250 ml to 500 ml	
less than 250 ml	

© HarperCollinsPublishers 2016

Issue 24 – Time
S&C Volume 3

The Maths Herald

Name:

Date:

The Language of Maths

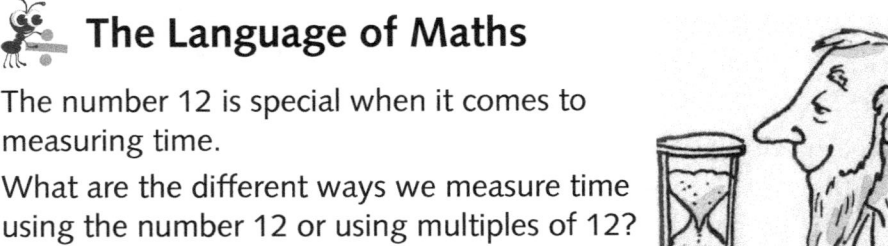

The number 12 is special when it comes to measuring time.

What are the different ways we measure time using the number 12 or using multiples of 12? What other numbers are important when measuring time?

Sports Update

Mrs Keft runs the Winston Hills Children's Netball Club. This Saturday she has to schedule 8 games on the club's 1 netball court. Each game lasts approximately 30 minutes. The first game cannot start before 9:00 a.m., and all games must be finished by 2:30 p.m.

Write Mrs Keft's schedule.

What's the Problem?

"Dad, you're 5 times as old as I am."

"That's right Oliver, but in 8 years time I'll only be 3 times as old as you."

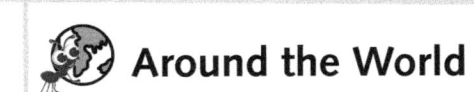

How old are Oliver and his father?

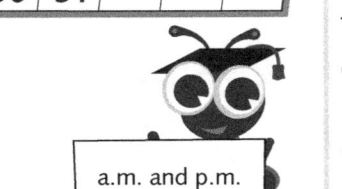

Around the World

Ferries leave Dover every hour, on the hour, to sail to Calais.

At the same time, ferries leave Calais every hour, on the hour, to sail to Dover.

The journey takes 2 hours in either direction.

How many ferries coming from Calais does a ferry leaving Dover pass as it crosses to Calais?

Focus on Science

How many months of the year have 31 days in them?

MARCH						
S	M	T	W	T	F	S
	1	2	3	4	5	6
7	8	9	10	11	12	13
14	15	16	17	18	19	20
21	22	23	24	25	26	27
28	29	30	31			

a.m. and p.m.

© HarperCollins Publishers 2016

Issue 24 — Time

Around the World

Plan a one-day educational visit for your class.

Things to think about:

- Where will you go? Why?
- How will you get there and back? How long will it take to travel there and back?
- How long do you need to spend there to make the visit worthwhile?
- What arrangements need to be made for food?
- Will you need to have a slightly longer school day? If so, why?

Write a detailed timetable for the day, starting from the moment you leave school until the time you return back.

Looking for Patterns

Cerqueto has a town clock that strikes once at 1 o'clock, twice at 2 o'clock, 3 times at 3 o'clock, and so on.

How many times does the clock strike in one full day?

How did you work out the answer?

The Puzzler

It takes 3 minutes for Simon the plumber to join together 2 pieces of piping.

How long will it take him to join together 8 pipes if he joins them all at the same speed?

What's the Problem?

Michael the electrician had 3 jobs today. He finished the last job of the day at 4:45 p.m. This job took him $2\frac{1}{2}$ hours to complete.

The job before this only took him 50 minutes, and the first job of the day took him $3\frac{1}{4}$ hours to complete.

Apart from working, Michael had a 1 hour lunch break and spent a total of 2 hours 10 minutes travelling.

At what time did Michael start work?

What's the Problem?

The Bike Borrower Company hires out bicycles.

They charge £8.00 for 6 hours or £6.00 for 4 hours.

If you are late in returning the bike, the company charges £1 for the first hour and £3 for every hour after that.

Helen originally hired the bike for 6 hours. She was late in returning the bike and was therefore charged a total of £15.

For how many hours was Helen late in returning the bike?

© HarperCollinsPublishers 2016

Issue 25 – Measurement
S&C Volume 3

The Maths Herald

Name:

Date:

🐭 What's the Problem?

The ages of Trevor and his grandfather added together total 70 years.
If Trevor's grandfather is 60 years older than Trevor, what is Trevor's age?

S	M	T	W	T	F	S
	1	2	3	4	5	6
7	8	9	10	11	12	13
14	15	16	17	18	19	20
21	22	23	24	25	26	27
28	29	30	31			

area

🐭 Let's Investigate

Look at this calendar.
A box has been drawn around 4 dates.
Add the 2 pairs of numbers that are diagonally opposite.
What do you notice about the 2 answers?
Repeat for other sets of 4 numbers.
Does this work for any month?
What other patterns do you notice?

© HarperCollinsPublishers 2016

Issue 25 – Measurement

🐭 The Puzzler

A farmer has a rectangular shaped farm made up of 2 square fields and 1 rectangular field.

He gives away the 2 square fields to his children and keeps the rectangular field for himself.

If the field he keeps is 300 m by 500 m, what are the largest measurements that the farm could have been before the farmer divided it up?

🐭 Let's Investigate

Look at the front page of a newspaper.
Work out how much space is given to different parts of the page:
- the name of the newspaper
- headlines
- stories
- pictures
- advertisements
- white space
- anything else

Is this the same for the front pages of different newspapers?

🐭 What's the Problem?

Roshan has to cross a desert on foot.
It is a 6-day journey to cross the desert. She is only able to carry 4 litres of water at a time, and she needs to drink 1 litre each day.

How can she cross the desert?

Issue 25 — Measurement

Let's Investigate

Estimate how many sheets of A4 paper it would take to cover your desk.

Write down your estimate.

Now measure as accurately as possible using sheets of paper.

Write down the result. How close was your estimate?

Use this answer to work out how many playing cards it would take to cover your desk.

Write down how you did this.

Let's Investigate

Make a list of all the rooms in your school – the classrooms, hall, storerooms, offices and any other rooms.

Now put them in order of *floor area*, from largest to smallest.

Compare your list with a friend's.

Do you agree on the order?

If you don't agree, how are you going to find out who is right?

Looking for Patterns

The rectangle on the right has an area of 36 square centimetres (written as 36 cm^2).

Using squared paper, draw as many different rectangles as you can that have an area of 36 cm^2.

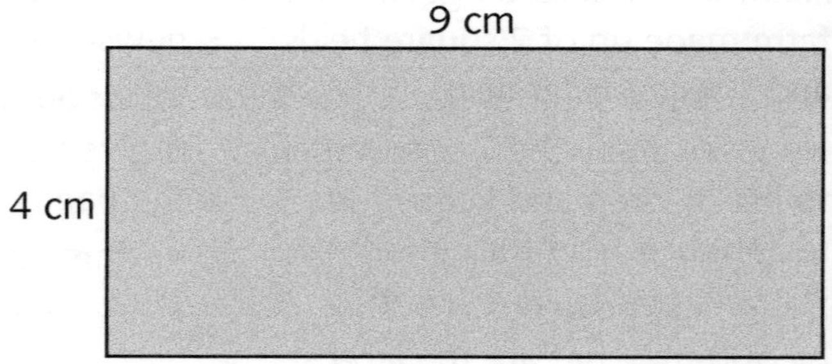

4 cm

9 cm

Let's Investigate

As in the Looking for Patterns activity above, using squared paper, draw different triangles all with areas of 12 cm^2.

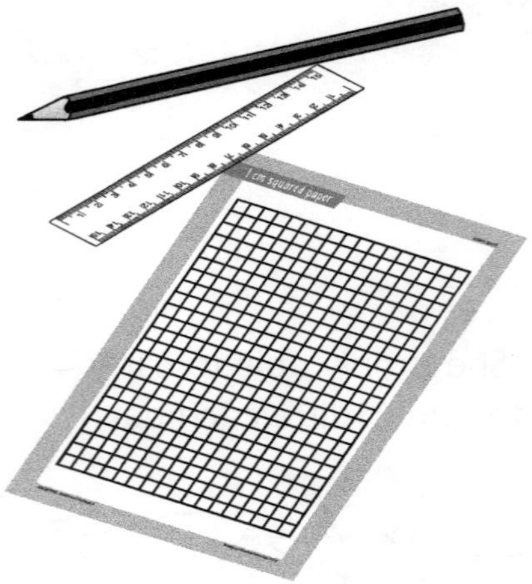

Construct

Make the largest envelope you can from a single sheet of A4 paper.

Write a set of instructions on how to make the envelope, including measurements and diagrams.

Issue 26 – Measurement

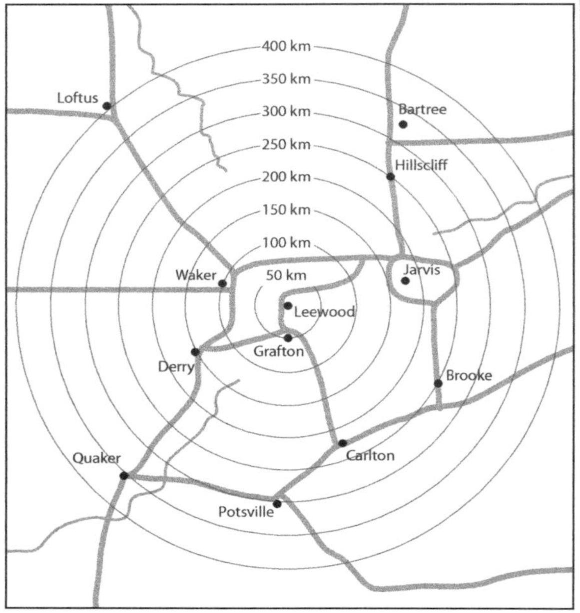

S&C Volume 3

The Maths Herald

Name:

Date:

Looking for Patterns

Mr Wilson, the school caretaker, has just finished putting a fence around the new nursery playground.

The playground is square in shape with each side measuring 8 m.

Mr Wilson used 20 fence posts to build the fence.

He used the same number of posts on each side of the fence.

How many posts are on each side of the fence?

perimeter

The Language of Maths

What maths words are used to described the weather?

Make a list of all the words.

Now look at the weather forecasts in different newspapers. Add any new words to your list.

What's the Problem?

The Sunny Pear Company makes a drink from pear juice and water.

The drink is mixed in a large container filled by 2 taps: one flows with pear juice and the other with water.

The pear juice tap can fill the container in 60 minutes; the water tap can fill the container in 30 minutes.

How long does it take to fill the container using both taps?

© HarperCollins*Publishers* 2016

The Language of Maths

A *scale* is a system of measurement based on a series of divisions laid down at regular intervals and representing numerical values.

A ruler and a measuring jug are 2 instruments that have a scale.

Make a list of all the instruments you can think of that use a scale.

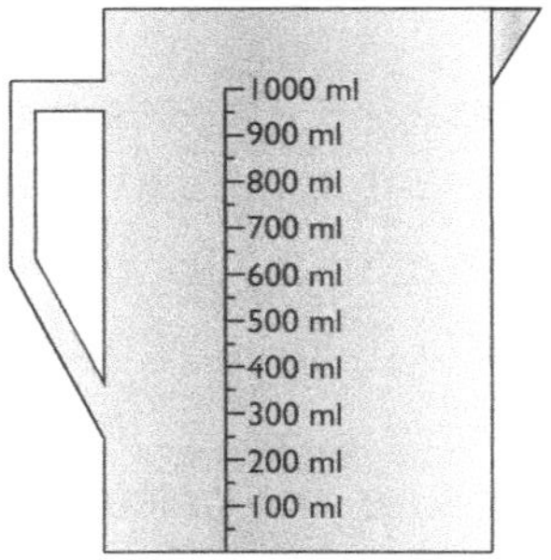

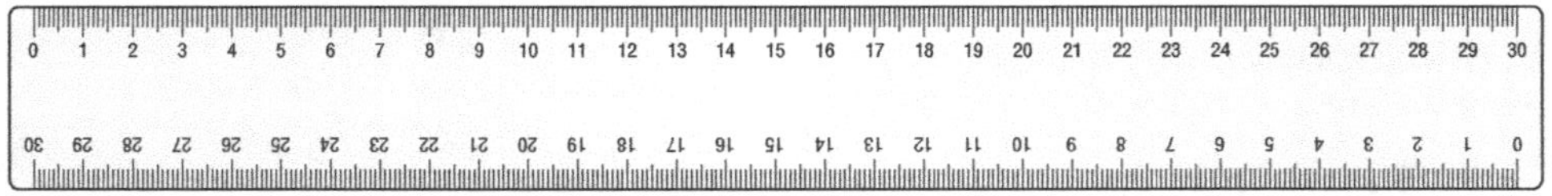

Around the World

On some maps, *concentric circles* (circles of different sizes with the same middle point) are used to calculate distances between places.

For example:

- The distance between Waker and Loftus is 300 km.
- The distance from Bartree to Brooke is approximately 150 km.

Write different statements about the distances between towns on the map.

Issue 26 — Measurement

Around the World

This activity requires you to think about distance and time.
Read the following statements.

> The child in our class who lives furthest from school takes the longest to get to school.

> The child in our class who lives closest to school is the quickest to get to school.

Think carefully about how you are going to find out if the answers to both these statements are true.
Write about what you do.
Write a statement justifying your conclusions.

What's the Problem?

The magic tree frog is very strange. It can only jump 2 distances: 40 cm or 70 cm, either forward or backward.

To reach an insect 110 cm ahead of it, it takes 2 jumps, one of 40 cm and one of 70 cm.

Assuming the tree frog only makes straight line jumps towards an insect, what is the fewest number of jumps it can take to reach an insect 90 cm in front of it?

Let's Investigate

Investigate the *perimeters* of different spaces at your school.
For example, your classroom, the school hall, playground, playing field, …
Measure and calculate the perimeter of each of these spaces.
Write about what you did.

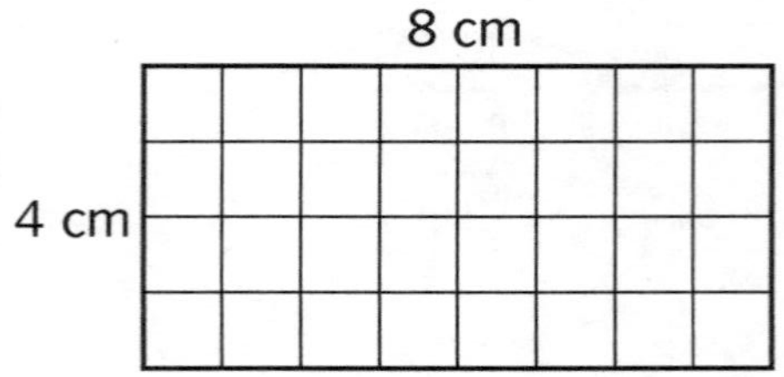

Focus on Science

An Earth year is 365.24 days and an Earth day is 24 hours.
On the imaginary planet Xavion a year is 96 days and a day is 8 hours.

Design a Xavion calendar, including the following:

- The number of months in the year
- The number of days in each month
- The number of days in each week

What are you going to call your months and days?

Looking for Patterns

The rectangle below has a perimeter of 24 cm.

Using squared paper, draw as many different rectangles as you can that have a perimeter of 24 cm.

4 cm

8 cm

© HarperCollins*Publishers* 2016

Issue 27 – 2-D shapes
S&C Volume 3

The Maths Herald

Name:　　　　　　　　　　Date:

The Language of Maths

Draw a design or picture using 8 shapes – a small and a large: square, rectangle, triangle and circle.

Imagine you had to explain what you have drawn to someone over the telephone.

Write a set of instructions telling them what to do.

Give your set of instructions to a friend. (Don't show them your diagram.)

Can they accurately draw your diagram?

The Puzzler

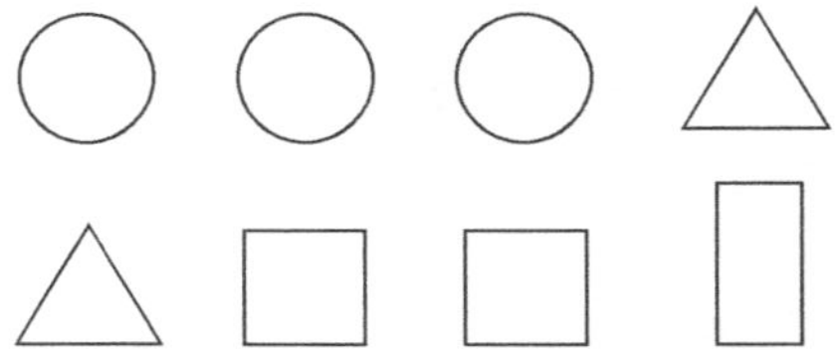

Use the clues to arrange the 8 shapes on the grid.

Clue 1: The centre of the bottom row has circles.

Clue 2: A circle is above a square.

Clue 3: The rectangle is between a circle and a square.

Clue 4: A square is in the bottom left corner.

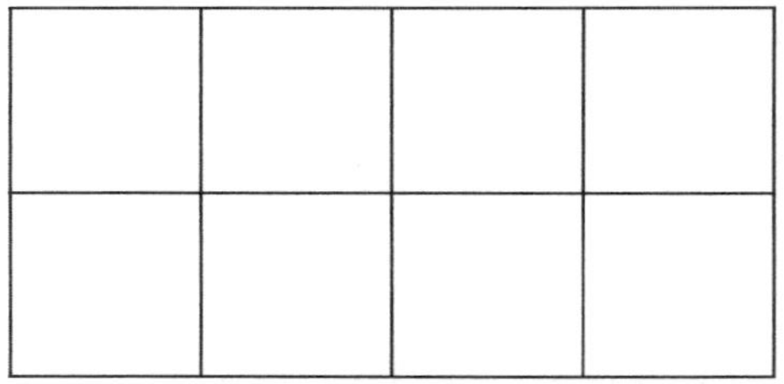

Let's Investigate

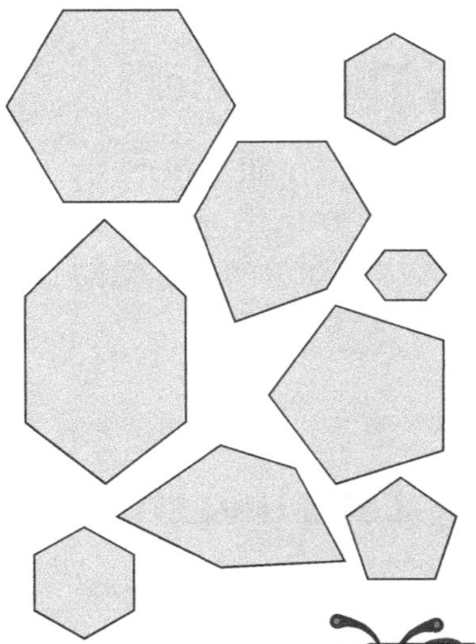

Draw some pentagons with right angles. What is the maximum number of right angles a pentagon can have?

Draw some hexagons with right angles. What is the maximum number of right angles a hexagon can have?

Try different polygons.

What conclusions can you make about the maximum number of right angles that different polygons can have?

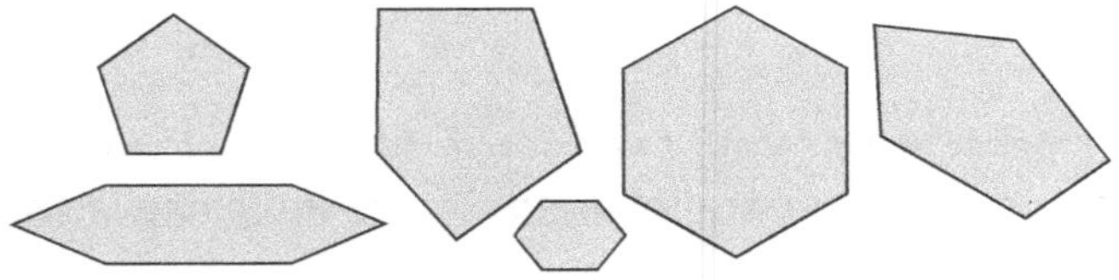

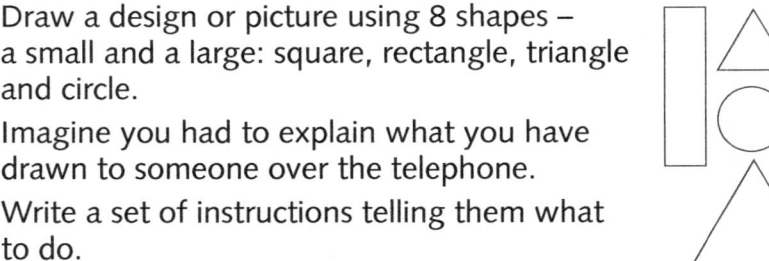

regular polygon

The Arts Roundup

Gillian did this painting. It was inspired by a Dutch painter called Piet Mondrian.

How many squares can you see?

How many rectangles?

Investigate pictures by Mondrian.

What is the greatest number of squares you can find in a picture by him?

© HarperCollins*Publishers* 2016

Issue 27 – 2-D shapes

The Puzzler

This 2 × 2 square has been made using 12 matchsticks.

- Move 3 matchsticks to make 3 squares all the same size.
- Move 4 matchsticks to make 3 squares all the same size.

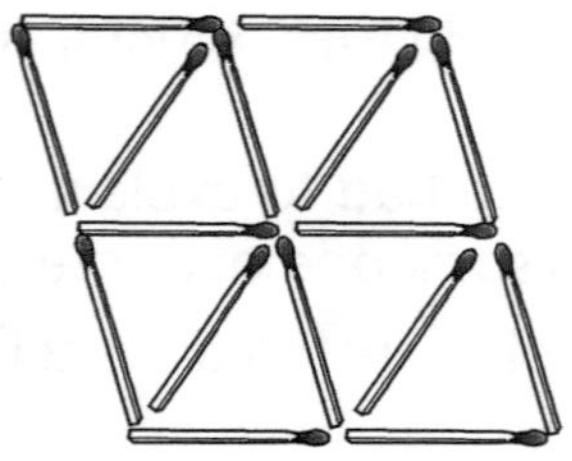

Draw your new squares and triangles.

This parallelogram has been made using 16 matchsticks.

- Remove 6 matchsticks to leave 2 triangles.
- Remove 4 matchsticks to leave 6 triangles.

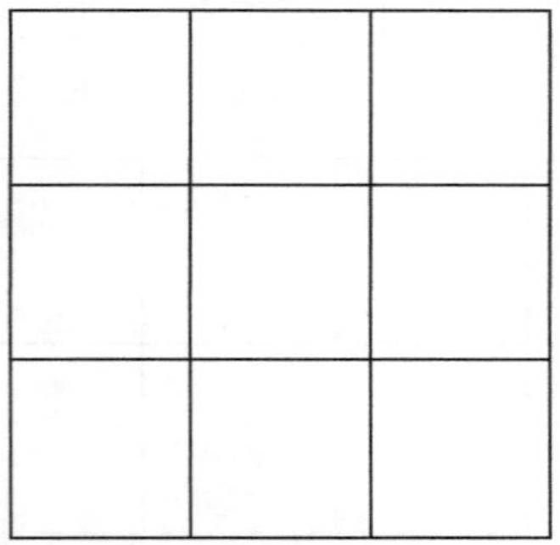

Looking for Patterns

What is the largest number of equilateral triangles can you make using 9 matchsticks?

How many squares are there here altogether?

The Puzzler

Can you join all 9 dots by drawing four straight lines without lifting your pencil off the paper?

Let's Investigate

Take a sheet of A4 paper.

Using a ruler, draw a pair of parallel lines across the paper from one edge to the other.

Draw 3 more straight lines each going in a different direction, from edge to edge, so that they cross the parallel lines. These lines can also cross each other if you want.

Cut out all the shapes these lines have made.

Now name and sort the shapes.

Can you sort them in more than one way?

© HarperCollins*Publishers* 2016

Issue 28 – 3-D shapes
S&C Volume 3

The Maths Herald

Name:

Date:

🐜 Construct

Imagine you had 120 blocks to make a pyramid with 1 block on the top, 2 blocks in the next row, 3 blocks in the next row, and so on.

How many blocks would be in the bottom row?

How many rows would there be?

🐜 Construct

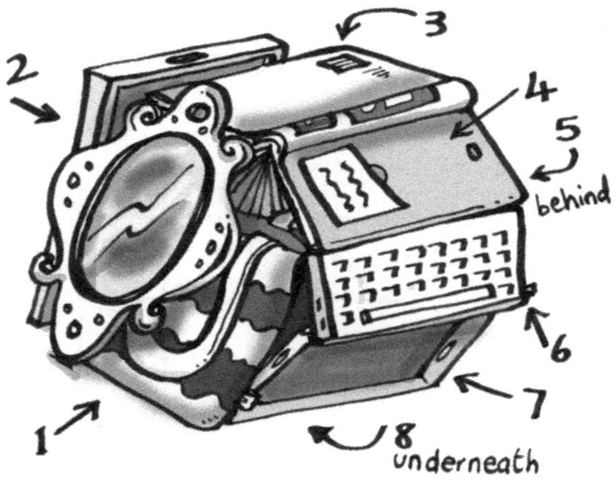

Find or make examples of solid shapes with 1, 2, 3, 4, 5, 6, 7 and 8 faces. You will need to make 8 shapes in total.

The shapes do not need to be regular.

Use junk modelling materials and be as imaginative as you can.

© HarperCollinsPublishers 2016

🐜 Construct

Using 27 interlocking cubes build a model using these 3 views as a guide.

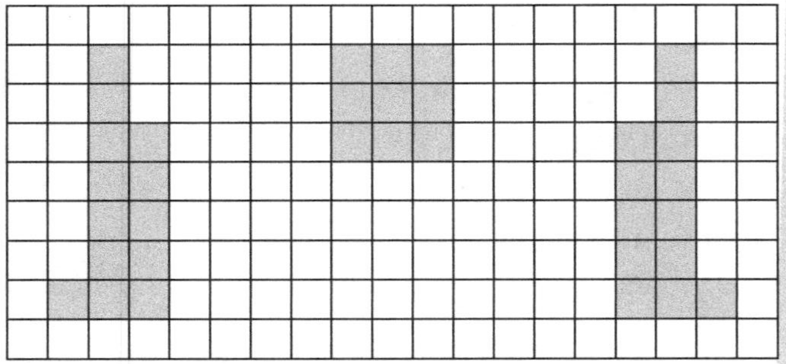

| Front view | Top view | Side view |

🐜 Construct

Using squared paper, draw the front view, top view and side view of this model.

🐜 Construct

Using between 10 and 30 interlocking cubes, secretly construct a model.

Use squared paper to draw the front view, top view and side view of your model.

Then give your three views to a friend. Can they make your model?

Issue 28 – 3-D shapes

At Home

In the supermarket, look at all the different shapes of packaging. Write down the names of all the different shapes you can see. Which are the most common shapes? Why do you think this is?

What's the Problem?

A cubic block of cheese is covered in rind.
It is cut up into 64 equal cubes.
How many of the smaller cubes have rind on 3 sides?

Construct

How many different ways can you colour the faces of a cube if 3 faces must be red and 3 faces must be blue?
Draw nets of a cube on squared paper to show your answers.

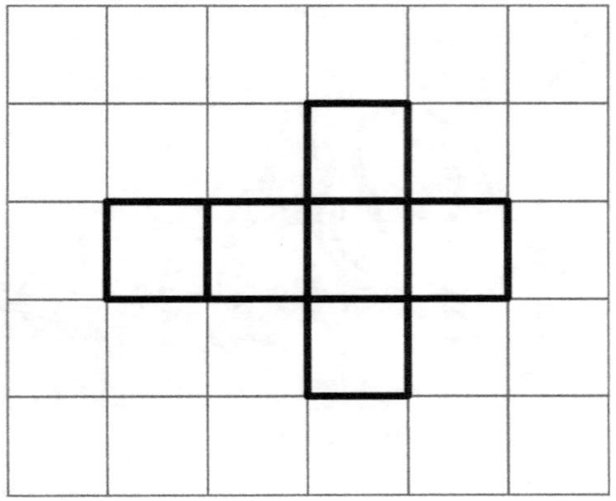

Construct

This cuboid has been made using 12 interlocking cubes.
It can be expressed as the following calculation:

$$3 \times 2 \times 2$$

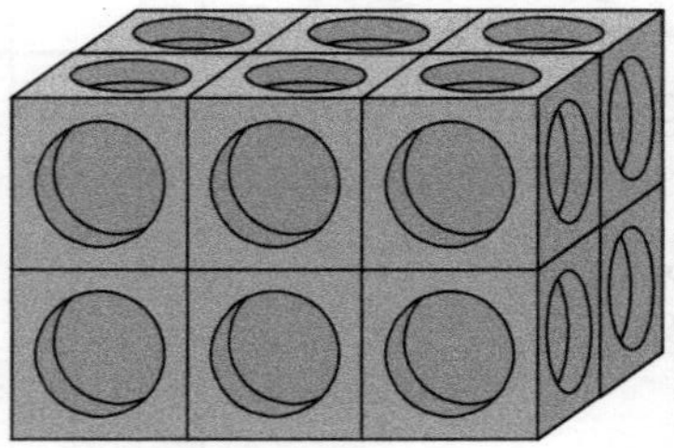

How many different cuboid shapes can you make with 24 interlocking cubes?
Write a calculation for each shape.

Construct

Look at the model made from cubes.
A picture of the model has been drawn using triangular dot paper.
Take 8 interlocking cubes and make your own model.
Now draw your model using triangular dot paper.

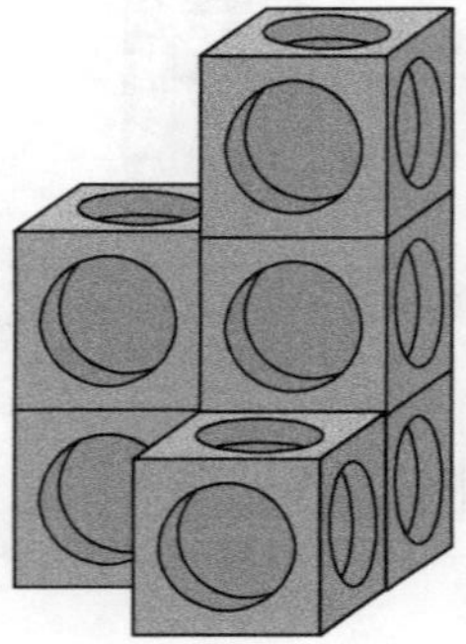

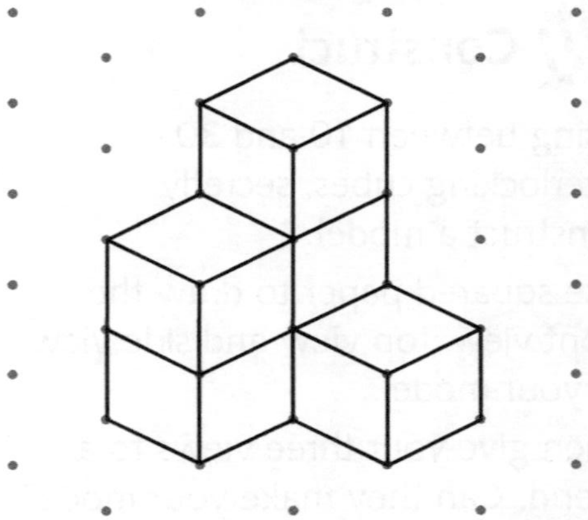

Issue 29 – Symmetry
S&C Volume 3

The Maths Herald

Name:

Date:

Construct

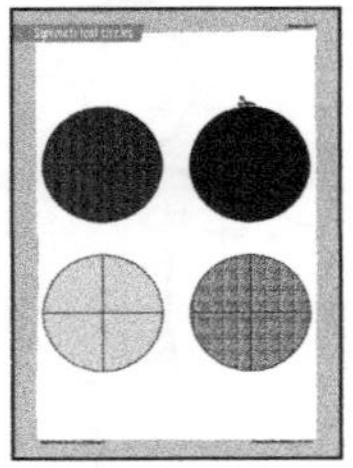

Using interlocking cubes, make different shapes that show reflective symmetry.

Think about using different coloured cubes to demonstrate the symmetry.

Try and make different shapes that show horizontal and vertical symmetry.

What about diagonal symmetry?

For each shape you make, write a statement describing the shape's symmetry.

Let's Investigate

Using the school camera, take photographs of any objects in your classroom and around the school that show reflective symmetry.

Make sure that you take the photos from an angle that shows reflective symmetry.

Print your photos and draw the lines of symmetry on them.

Do any of your objects have more than one line of reflective symmetry?

© HarperCollinsPublishers 2016

Construct

Cut out the 4 circles on the Resource sheet.

Then cut each of these circles into quarters.

Now using some or all of the 16 quarters, place them onto squared paper to make a symmetrical pattern. Your pattern can have a vertical, a horizontal or a diagonal line of symmetry.

What different patterns can you make?

Vertical line of symmetry

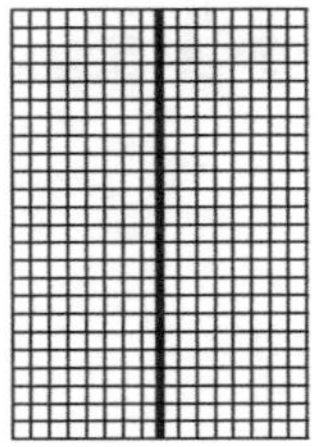

Horizontal line of symmetry

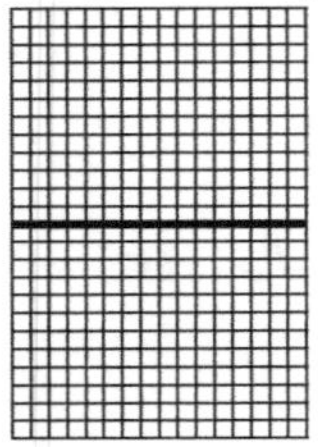

Diagonal line of symmetry

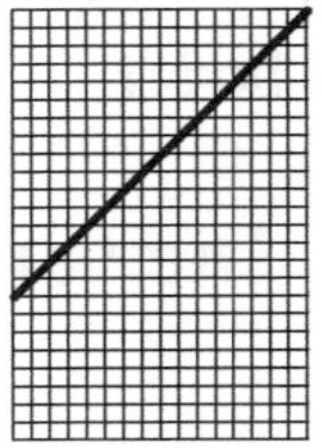

Glue your best patterns onto squared paper.

Looking for Patterns

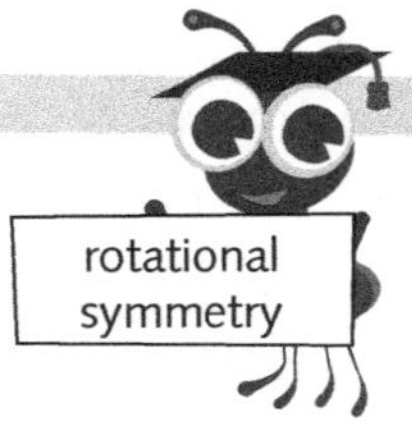

Look at the grids below. Moving from Grid 1 to Grid 3, the pattern of shaded squares changes.

Identify the pattern and colour Grids 4 and 5 following this pattern.

Grid 1   Grid 2   Grid 3   Grid 4   Grid 5

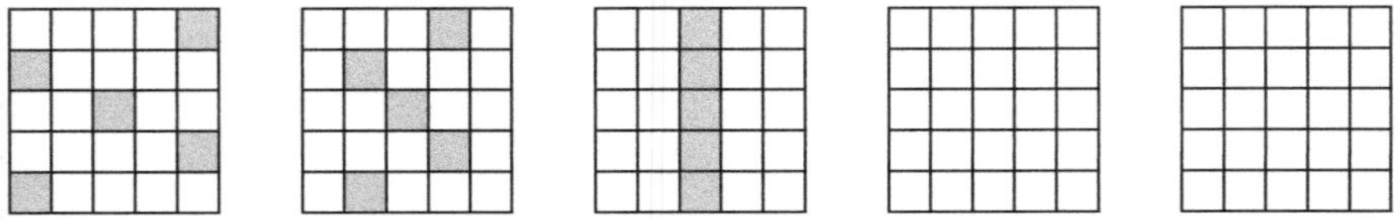

Look at the 5 completed grids. Which pairs of grids, when placed side-by-side, show reflective symmetry?

Draw these pairs of grids on squared paper.

rotational symmetry

Grid ?   Grid ?

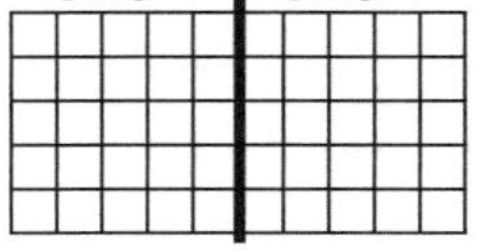

Grid ?   Grid ?

Issue 29 – Symmetry

Famous Mathematicians

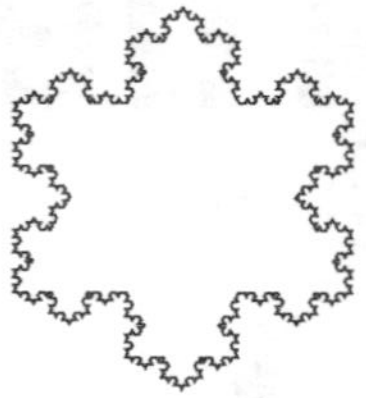

The *Koch Snowflake*, also known as the *Koch Star* and *Koch Island*, is a mathematical pattern named after a Swedish mathematician by the name of Helge von Koch.

To construct a Koch Snowflake begin with an equilateral triangle. For each side of the triangle, divide the side into thirds. On the middle third, draw an equilateral triangle.

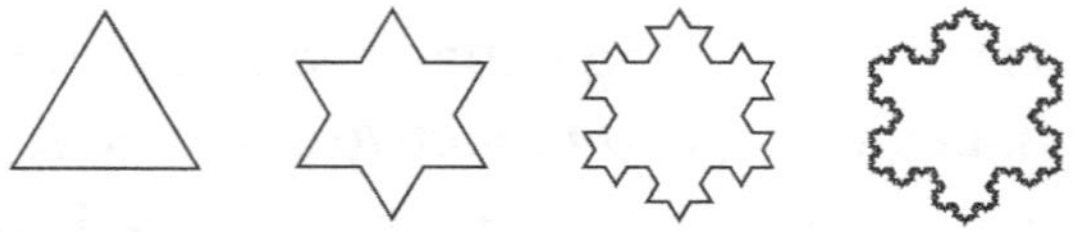

When complete, repeat this process for each new triangle.

The Koch Snowflake on the right has been constructed using triangular dot paper.

Use a sheet of triangular dot paper to construct your own Koch Snowflake.

Now use different colours to show reflective symmetry.

Write a report on your Koch Snowflake commenting on its symmetry.

Let's Investigate

Is your face symmetrical?
Are your ears level with your eyes?
Are your ear lobes level with each other?
Are your eyes halfway between the top and bottom of your head?
Measuring from side to side, is your nose in the middle of your face?
Is your smile symmetrical?
How will you find out the answers to these questions?
What else can you find out about the symmetry of your face?

Issue 29 – Symmetry

Let's Investigate

Is your body symmetrical?
Are your feet the same size?
Is your navel in the middle of your body?
Are your knees at the same level?
Are your arms the same length?
How will you find out the answers to these questions?
What else can you find out about the symmetry of your body?

The Language of Maths
Where do fish keep their money?

Draw the mirror image of the route in the left-hand grid onto the right-hand grid.

Note each number where your route has changed direction in the right-hand grid, and enter it in the top row of the grid below.

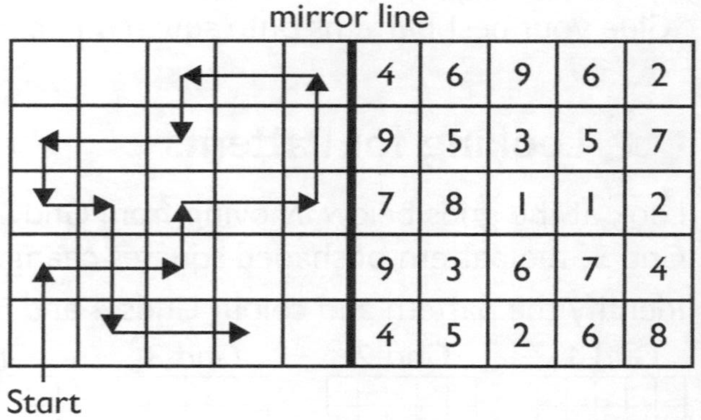

Numbers		
Letters		

Now use the code to find the answer to the riddle.

1	2	3	4	5	6	7	8	9
A	B	E	I	K	N	R	T	V

© HarperCollins*Publishers* 2016

The Maths Herald

S&C Volume 3

Name: _____ Date: _____

The Language of Maths

Hide some treasure in the classroom, and then make a map to show where it is hidden.

Now hide the map, and make up some clues about where to find the map.

The Puzzler

Read the clues to colour the stars and solve the puzzle.

You need a red, blue, green, yellow, black and purple coloured pencil.

Clue 1: The blue star is between the red star and the white star.
Clue 2: There is 1 star between the red star and a green star.
Clue 3: The purple star is between the yellow star and a green star.
Clue 4: A green star is adjacent to and anticlockwise from the uppermost star.
Clue 5: The two green stars are adjacent.
Clue 6: The uppermost star is neither green nor purple.

© HarperCollins*Publishers* 2016

The Puzzler

Anna and 3 friends are sitting around a table playing a game. They are using a yellow, a red, a green and a blue counter.

Cathy sits on the left of Bashir and opposite Dennis.

Anna uses a red counter. Dennis isn't using a blue counter and the person using a yellow counter is sitting on Anna's right.

What colour counter is each person using?

____° degrees

The Language of Maths

What makes an octopus laugh?

Follow the directions to find the answer to the riddle. Write each letter you land on in the boxes below.

- Go up 3 squares.
- Go right 3 squares.
- Go up 2 squares.
- Go left 1 square.
- Go down 4 squares.
- Go left 1 square.
- Go up 3 squares.
- Go right 3 squares.
- Go down 2 squares.
- Go left 4 squares.

G	H	T	N	O
V	K	B	C	L
T	M	D	E	U
S	P	Y	K	E
A	C	I	F	R

Start ↑

Issue 30 — Position and direction

Technology Today

This floor robot was given the following program:

- Forward 3 spaces.
- Turn left 90°.
- Forward 1 space.
- Turn left 90°.
- Forward 1 space.
- Turn left 90°.

Using squared dot paper, the program can be shown as:

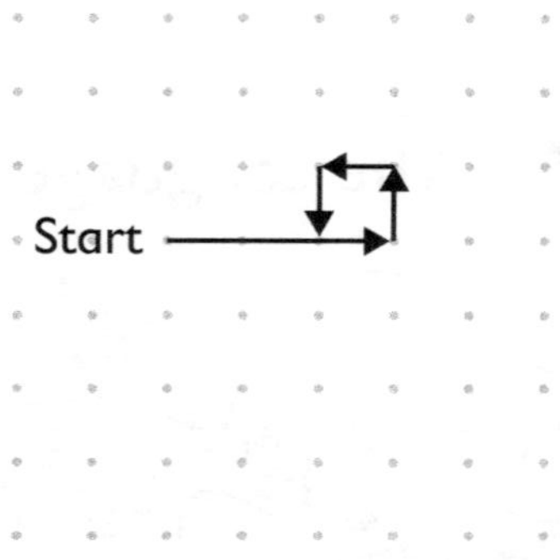

Start →

The robot kept on repeating the pattern. Continue the robot's route.

Technology Today

This floor robot was given this program:

- Forward 1 space.
- Turn right 90°.
- Forward 2 spaces.
- Turn right 90°.
- Forward 4 spaces.
- Turn right 90°.

Using squared dot paper, show the robot's route.

Write a different path for the robot.

What's the Problem?

Jason, Helen and Louise all live in Alton Street. There are 8 houses between Jason's home and Louise's home, and there are 3 houses between Louise's home and Helen's home. How many houses are between Jason's home and Helen's home?

At Home

Many religious places of worship face in a particular direction.

Make a simple drawing of a church, mosque or temple near to where you live.

Can you work out which direction it faces?

Why do you think this is so?

Mark on your drawing the direction it faces.

Looking for Patterns

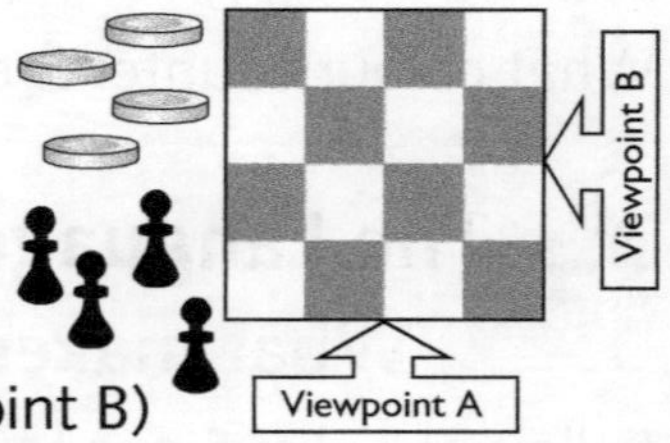

Look at the 4 × 4 draughts board, the four draughts pieces and four chess pieces.

Each of the arrangements below, show two different viewpoints (Viewpoint A and Viewpoint B) of how the 8 pieces are arranged on the board.

For each arrangement, each row contains one draughts and one chess piece, and each column also contains one draughts and one chess piece.

Imagine viewing the draughts board from both viewpoints on each of these arrangements. Using the Resource sheet, show how the pieces have been placed on each board.

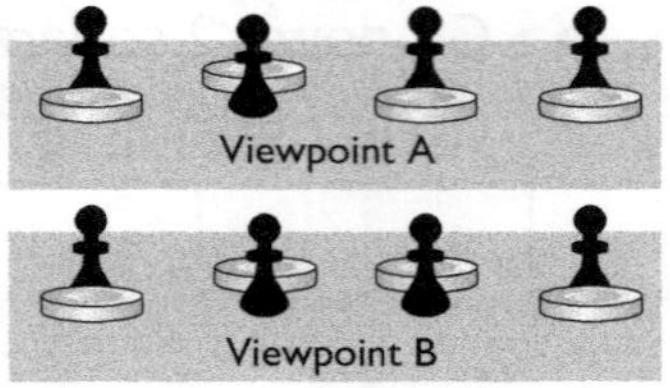

① Viewpoint A
② Viewpoint A
③ Viewpoint B
④ Viewpoint B

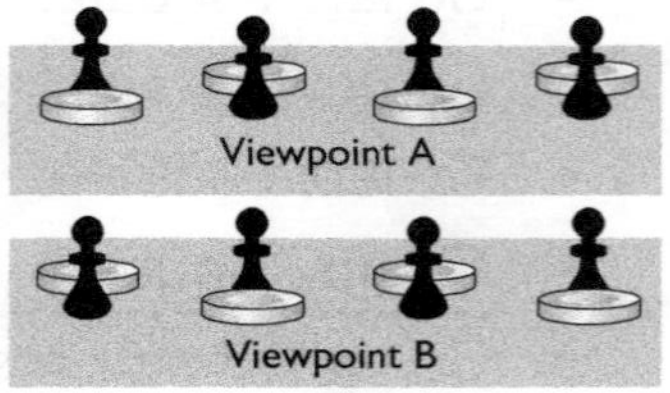

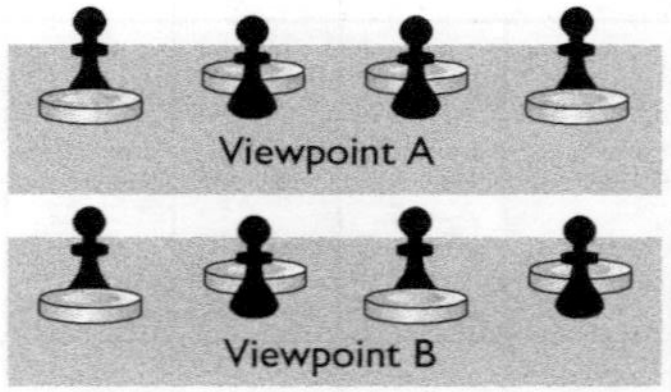

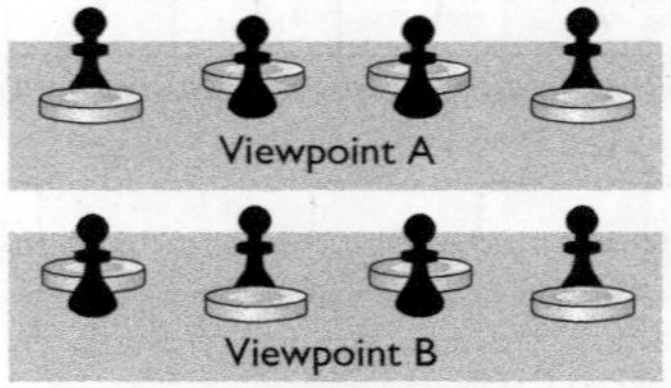

© HarperCollins*Publishers* 2016

The Maths Herald

S&C Volume 3

Name:
Date:

At Home

How many things do you "open" and "close" every day in your home? Make a list.

What about things you "press", "turn", "slide", "move up", "move down", "move in", "move out", ...

Focus on Science

The windmill on the right has 3 white sails and 1 black sail.
In which position A, B, C or D would the black sail be after the following turns?

1. 3 quarter turns clockwise
2. 4 quarter turns clockwise and 1 half turn anticlockwise
3. $\frac{1}{4}$ turn to the left
4. 2 right angles to the right
5. 6 right angles to the right and 1 right angle anticlockwise
6. 2 half turns anticlockwise and 1 half turn clockwise

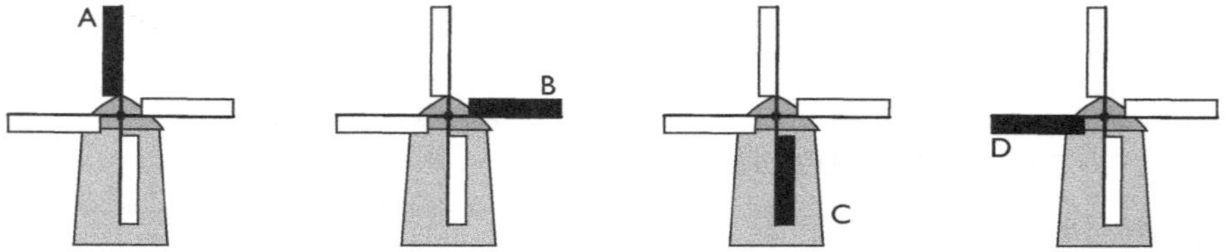

Around the World

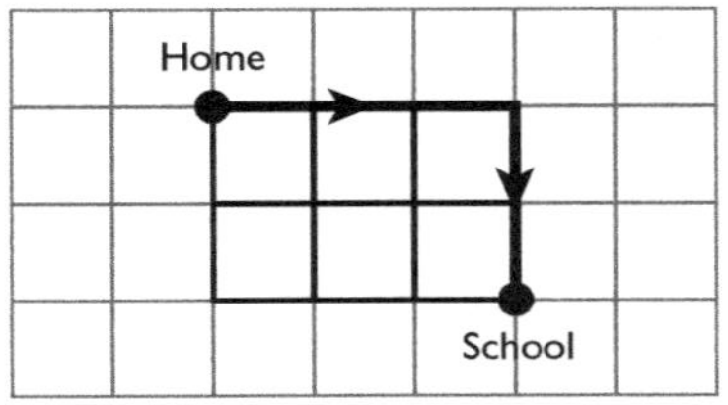

The picture below shows Ben's neighbourhood.

Ben and his mother like to take a different route from home to school each morning.

The diagram above shows one of the possible routes that Ben and his mother can take.

How many different routes can you find that are no longer than this one?

Draw your routes on squared paper.

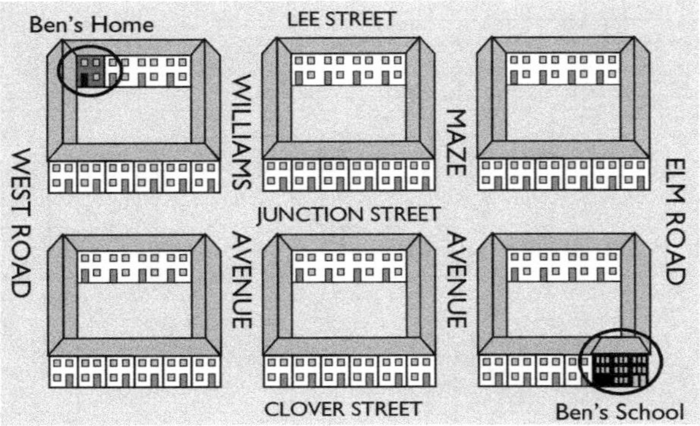

The Language of Maths

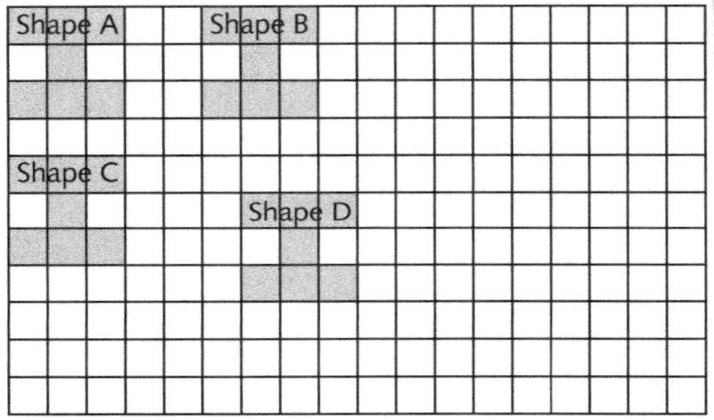

Shape A has been moved or *translated* 5 squares to the right to form Shape B.
Shape A has been *translated* 4 squares down to form Shape C.
Shape A has been *translated* 6 squares to the right and 5 squares down to form Shape D.

Translate Shape A:
- 14 squares to the right (label this Shape E)
- 8 squares down (label this Shape F)
- 11 squares to the right and 6 squares down (label this Shape G)

© HarperCollinsPublishers 2016

Issue 31 – Movement and angle

The Puzzler

A *maze* is an area of interconnected paths that it is hard to find a way through. They are usually designed in a garden with hedges between the paths.

Follow these instructions to get through the maze.

Draw your route on the maze as you go.

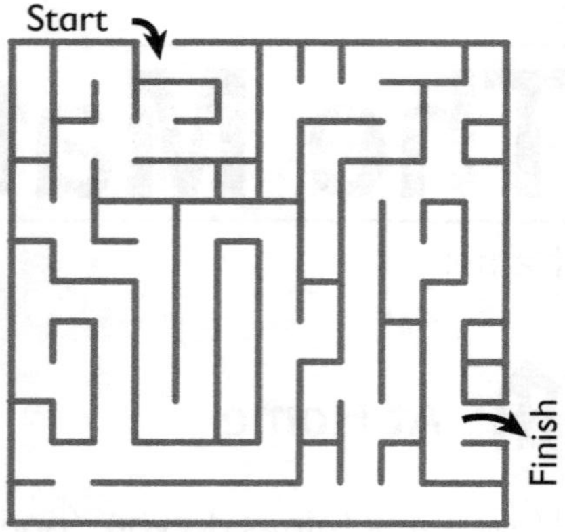

"Go through the entrance."
"Turn anticlockwise 90° and walk to the end."
"Turn clockwise 90° and walk to the end."
"Turn clockwise 90° and walk to the end."
"Turn anticlockwise 90° and walk to the end."
"Turn anticlockwise 90° and walk to the end."
"Turn clockwise 90° and walk to the end."

"Turn anticlockwise 90° and walk to the end."
"Turn clockwise 90° and walk to the end."
"Turn clockwise 90° and walk to the end."
"Turn anticlockwise 90°, walk forward and go through the first entrance on the left."
"Turn anticlockwise 90° and walk forward until you reach the second entrance on the left."

"Turn anticlockwise 90° and walk to the end."
"Turn clockwise 90° and walk to the end."
"Turn clockwise 90° and walk to the end."
"Turn anticlockwise 90°, walk forward until you reach the first entrance on the left."
"Turn anticlockwise 90° and walk forwards until you are out of the maze."

The Language of Maths

Find your way out of the maze.

Now draw your route on the maze.

Then write instructions for a friend to use to get out the same way.

Use terms similar to those above.

Now give your instructions to a friend to follow.

Do they take the same route as you? If not, why not?

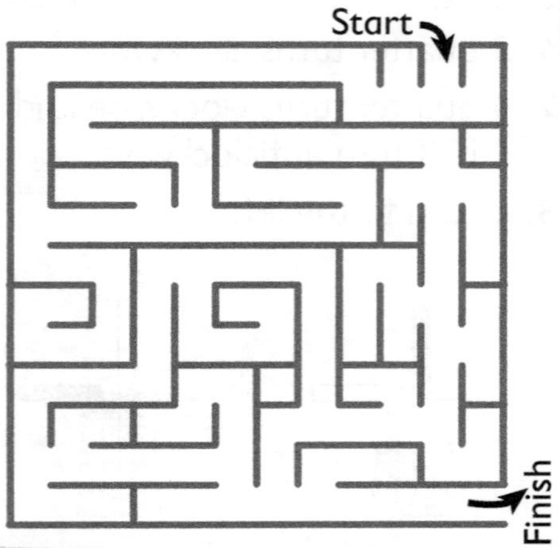

Looking for Patterns

Look at these 3 rows of solids.

In each row, the shapes have been turned clockwise 90°.

Show where the star is on each new shape.

translation

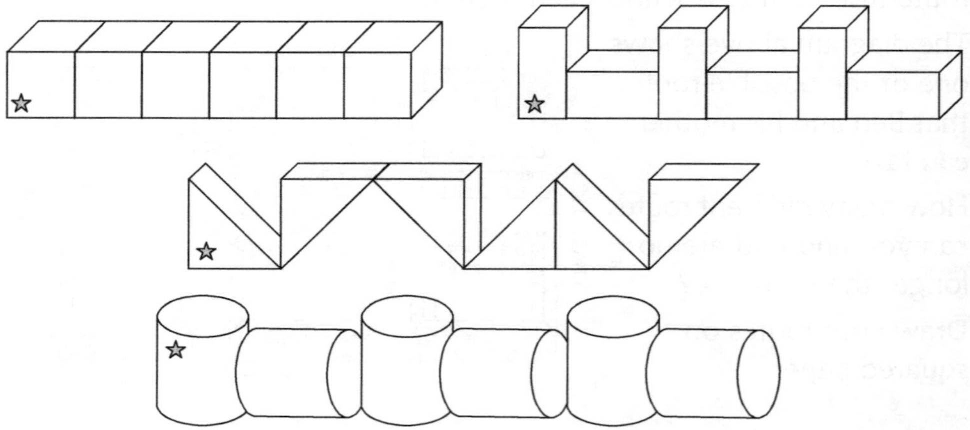

Looking for Patterns

Look at the pattern on the right.

Each time the line meets the side of the grid it is reflected 90° and continues.

Continue the pattern.

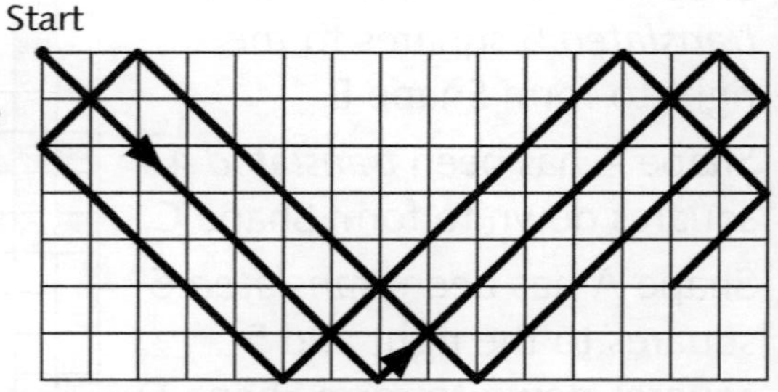

Using a sheet of squared paper, draw a large rectangle different from the one above. Start anywhere on the side of the grid and draw lines using the same method to make an interesting pattern.

What would happen if you started with a square?

© HarperCollins*Publishers* 2016

Issue 32 – Geometry

S&C Volume 3

The Maths Herald

Name:

Date:

Construct

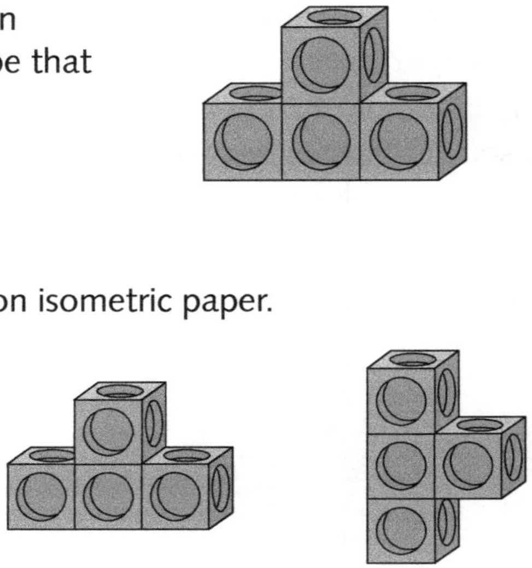

Using 4 interlocking cubes, join them together to make a shape that is 1 cube deep.

There are 5 different 4-cube shapes that you can make.

Can you find them all?

Record your different shapes on isometric paper.

Note: These 2 shapes are considered identical. One is simply the rotation of the other.

Looking for Patterns

Look at the 5 different 4-cubed shapes you made in the Construct activity above.

Which shapes, when duplicated 4 times, will fit together to make a 4 × 4 square?

Record your results on isometric paper.

The Puzzler

A large group of friends go out together for a meal in a restaurant. The restaurant is able to seat them round one large circular table.

After the waiter had taken everyone's order, he counted the number of friends sitting round the table to make sure that he had written down all the orders.

The waiter noticed that everyone was sitting opposite someone else across the table, and that the 3rd person and the 11th person were sitting directly opposite each other.

How many friends are sitting round the table?

Write about how you worked out the answer.

Looking for Patterns

5 interlocking cubes can be joined to build a 1-car garage, 6 interlocking cubes will give a 2-car garage, 7 interlocking cubes a 3-car garage, and so on.

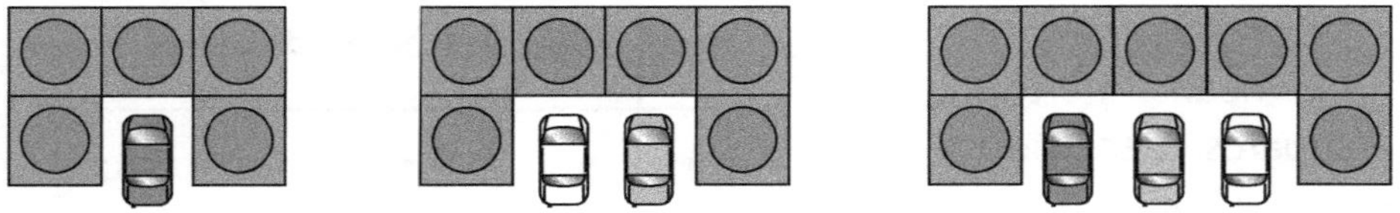

How many cars will go into a garage made of 12 cubes?

How many cubes are needed to make a garage that will fit 14 cars?

© HarperCollins*Publishers* 2016

Issue 32 — Geometry

Let's Investigate

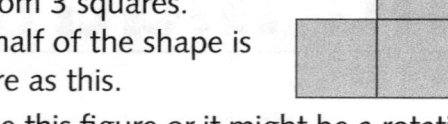

The figure on the right is made from 3 squares. It is only half a shape. The other half of the shape is made from exactly the same figure as this.

The other half might be exactly like this figure or it might be a rotation.

When joined together, at least 1 side of 1 square on each shape must be touching each other.

What different shapes can you make using 2 of the above figures in each shape?

Use the figures in Section A of the Resource sheet to help you construct your shapes.

Draw your shapes on squared paper.

What if these figures were half of the shape?

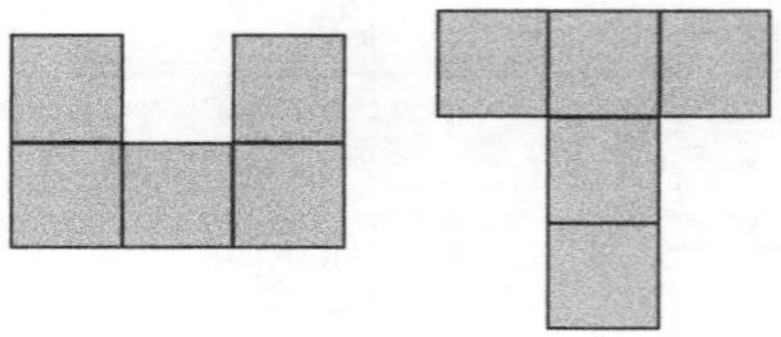

Looking for Patterns

How many triangles are there here altogether?

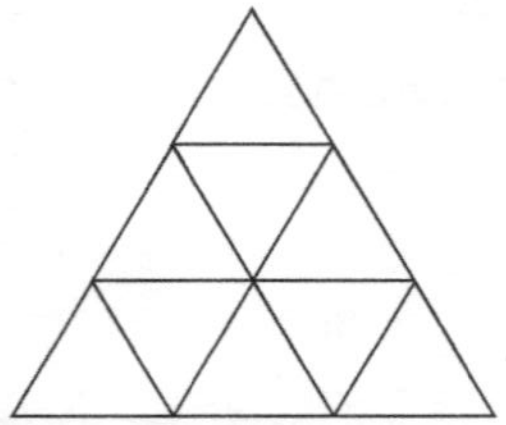

Let's Investigate

Two straight lines can be drawn in the following ways:

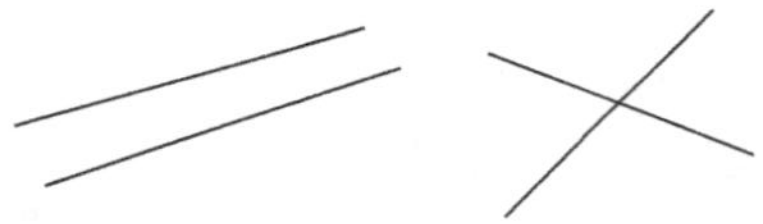

Investigate the different ways you can draw 3 straight lines.

The diagrams that have 2 straight lines have the following intersecting points:

Show the intersecting points for each diagram drawn with 3 straight lines.

The Puzzler

Can you tell which of these pieces fit together to make this regular hexagon?

The pieces can be rotated but not flipped.

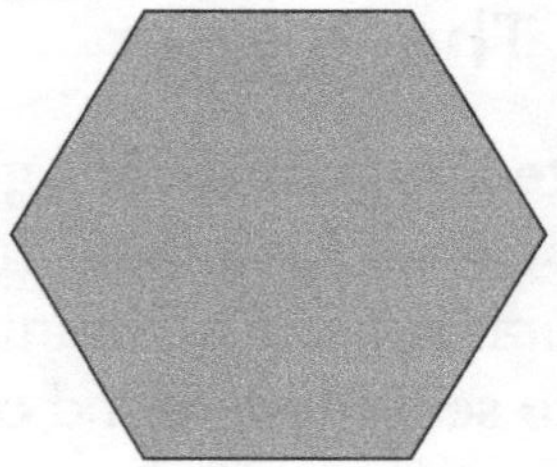

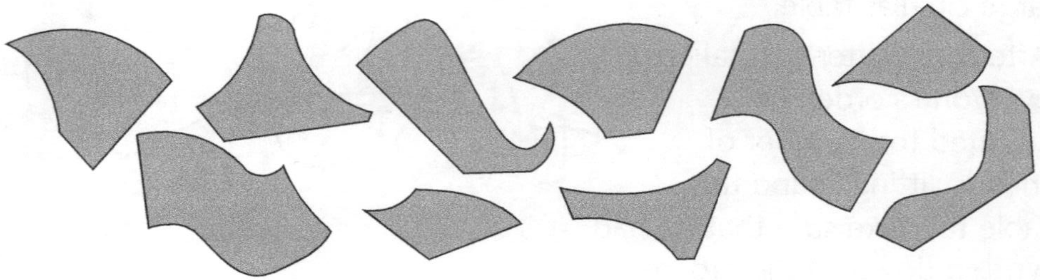

The Puzzler

You are approaching this island by boat. There is a chest of treasure in the centre of the island.

Follow the directions in each square to get to it.

For example:
- 3W means "go three squares to the west"
- 2N means "go two squares to the north".

On which square on the edge of the island do you need to land to reach the treasure?

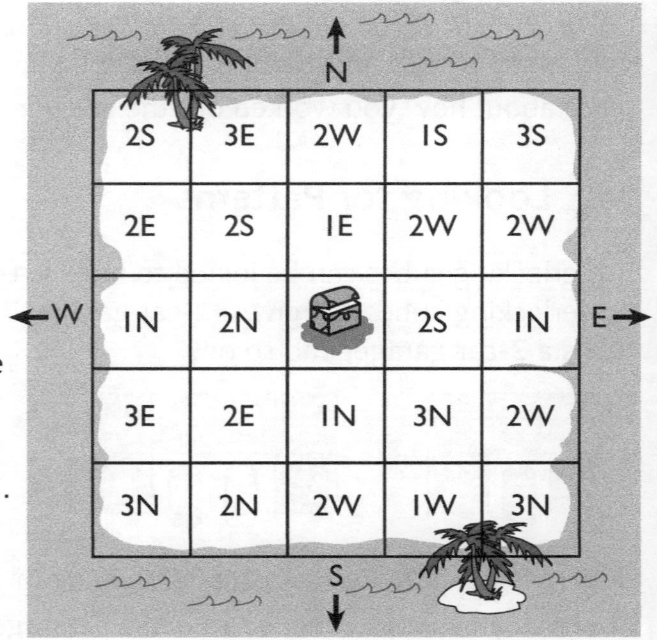

2S	3E	1E	2W	3S
2E	2S	2W	2W	1N
1N	2N		3N	2W
3E	2E	1N	2W	3N
3N	2N	2W	1W	3N

© HarperCollins*Publishers* 2016

Issue 33 – Geometry
S&C Volume 3

The Maths Herald

Name: Date:

The Puzzler

Complete the grid so that each row and column, and each 2 × 3 box, contains the 6 different shapes.

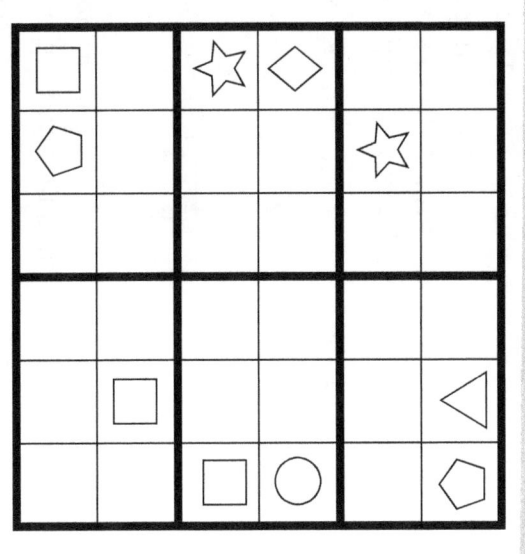

Construct

A *tromino* is a shape made from 3 identical squares.

There are 2 possible shapes:

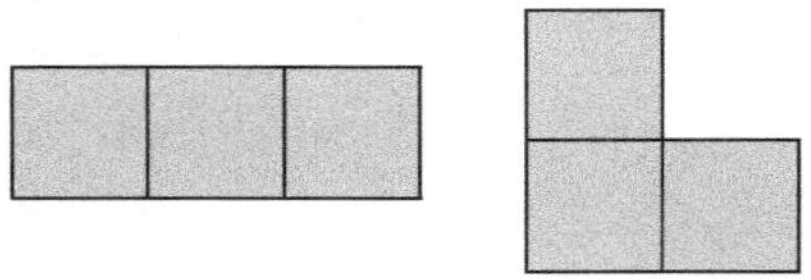

How many ways can you make a 2 × 3 rectangle using one or both of these shapes? What about a 3 × 3 square and a 3 × 4 rectangle? Use squared paper to record your results.

Hi! I'm a tromino. You probably know my cousin the domino.

© HarperCollins*Publishers* 2016

In the Past

The *tangram* was invented in China more than 4000 years ago. According to legend, a man called Tan was taking a ceramic tile to the Emperor when he slipped and dropped it. The tile broke into 7 pieces. While he was trying to put the tile back together, he found he could make many different figures and designs.

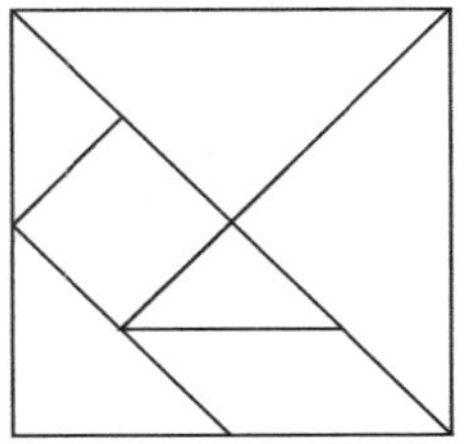

Cut out the tangram from the Resource sheet. Make each of the following shapes. No pieces can overlap and you must use all 7 pieces in each shape.

• triangle • rectangle • parallelogram • hexagon

The Puzzler

A Greek cross is made from 5 identical squares.

- Using squared paper, draw a Greek cross with the sides of each square 4 cm long.
- Draw a line through the cross as shown.
- Cut out your cross.
- Draw another line, perpendicular to the line you drew, so that your cross is divided into 4 pieces.
- Cut your cross into the 4 pieces along the pair of perpendicular lines.

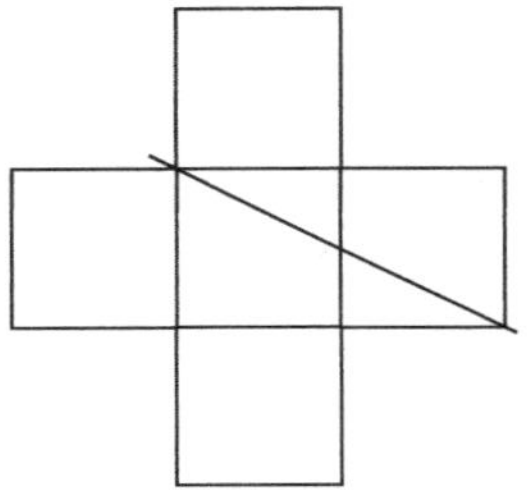

Can you rearrange the pieces to make a square?

Now draw another Greek cross with a line on it as shown, and cut it out.

- Draw a perpendicular line in a different place to last time.
- Cut your cross into pieces along the 2 perpendicular lines.

Can you rearrange the pieces to make a square this time?

Construct

Origami is the Japanese art of paper folding.
Follow these instructions to make a rectangular box.
You will need a sheet of coloured A4 paper.

Step 1
Fold the paper in half, then open.

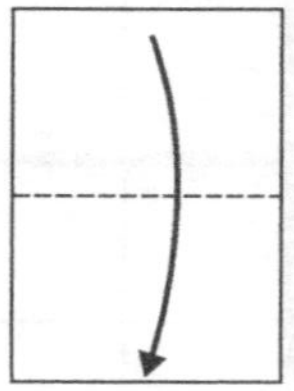

Step 2
Fold the top and bottom edges to the centre fold.

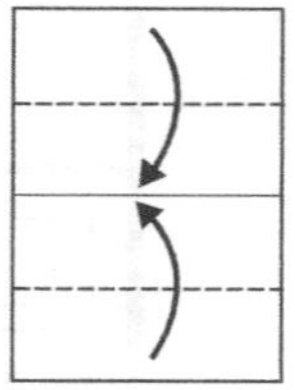

Step 3
Fold the paper in half, then open.

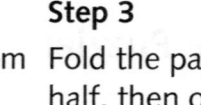

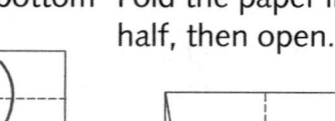

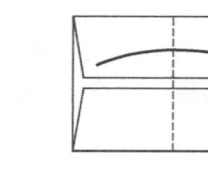

Step 4
Fold the left and right edges to the centre fold, then open.

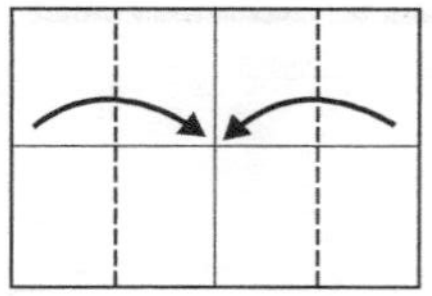

Step 5
Fold all four corners to the creases made in Step 4.

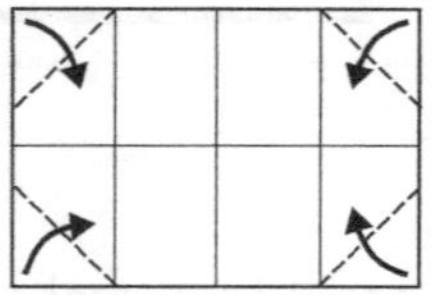

Step 6
Fold the inside flaps outwards to overlap the corners of the triangles.

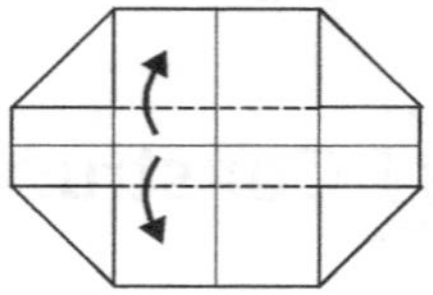

Step 7
Hold the model at the two points shown and open outwards. The box will take shape. You may need to crease the corners to make it more rectangular.

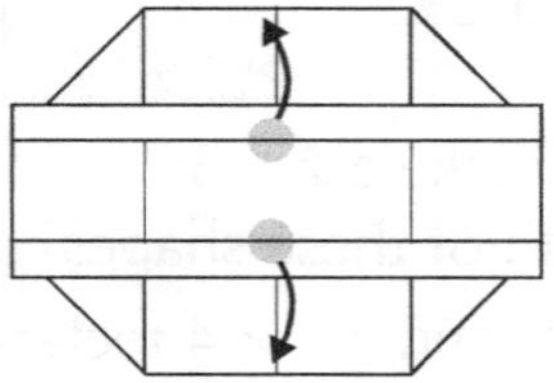

Step 8
Your finished box should look like this.

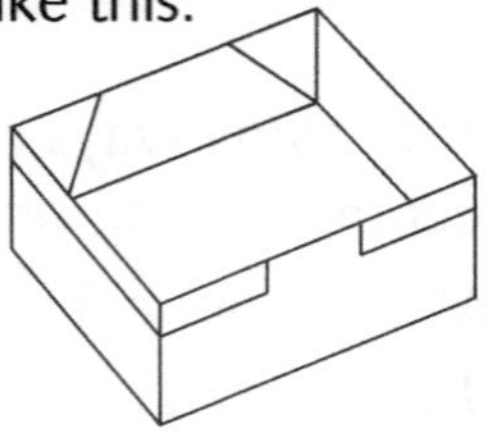

Looking for Patterns

Here are 4 different views of the same cube.

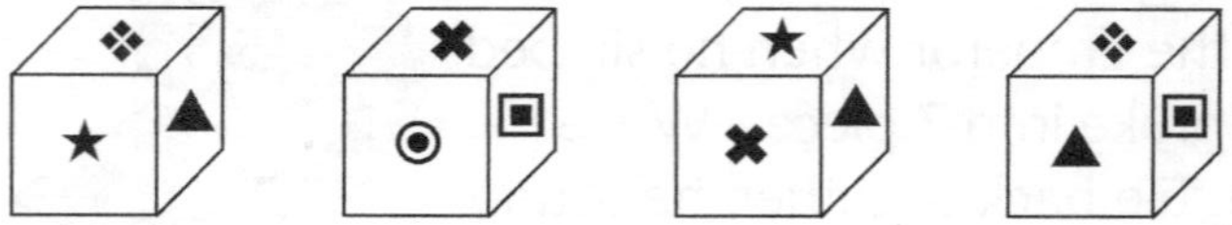

- What shape is on the opposite face to ■?
- What shape is on the opposite face to ▲?
- What shape is on the opposite face to ✖?

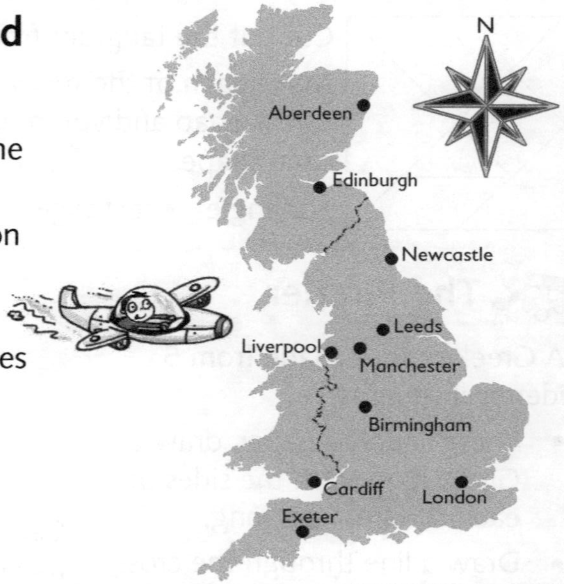

Around the World

Look at the map.
Imagine you own an aeroplane and you are the pilot!
If you were to fly from London to Cardiff, you would fly in a westerly direction.
Choose different pairs of places on the map and write about which direction you would travel if you flew from one place to the other.

Look at the map in the **Around the World** activity above. Either:

- mark on the map the town in which you live (that's if it can be found on the map or isn't marked on it already)
- mark another town or city that you know something about.

Write down which direction you would travel if you flew your plane from this town or city to some of the other cities marked on the map.

Issue 34 – Statistics
S&C Volume 3

The Maths Herald

Name: 	Date:

Famous Mathematicians

John Venn was a British mathematician who lived from 1834 to 1923. Venn was extremely interested in logic and probability. He created the sorting diagram that bears his name.

Draw Venn diagrams to show each of the following sets of information. Then answer the related questions.

1. 13 children have a brother.
   5 children have a brother and a sister.
   9 children have a sister.

   a. How many children have a brother but no sister?
   b. How many children have a sister but no brother?

2. 21 children like football.
   15 children like tennis.
   9 children like football and tennis.
   2 children do not like tennis or football.

   a. How many children like football and don't like tennis?
   b. How many children like tennis and don't like football?

Around the World

Choose a tourist attraction in your local area and collect as much information as you can about it.

Make a brochure about it. Be sure to include things such as prices and opening hours.

Include on your brochure a map to show visitors how to reach the attraction.

© HarperCollins Publishers 2016

Let's Investigate

Investigate who in your class has the largest extended family.

For the purposes of the investigation, an extended family consists of:

- parents
- children
- grandparents
- aunts and uncles
- cousins

How many extended family members are there in the "average" family in your class?

At Home

Which room in your home is used the most?

Which room is used the least?

Over the course of one evening, keep a record of the amount of time each member of your family spends in each room in your home.

Your record does not have to account for every minute, just use approximations.

Let's Investigate

Many of your characteristics are inherited from your parents.

One thing that is inherited is the ability to roll your tongue. Another is whether your ear lobes hang down from your ears or are attached to your head.

Do a survey of the children in your class to find out:

- how many can roll their tongues and how many cannot
- how many have hanging earlobes and how many have attached ones.

What fraction of children can roll their tongue?
Write a statement comparing the number of children that have hanging earlobes compared to those that have attached earlobes.

Let's Investigate

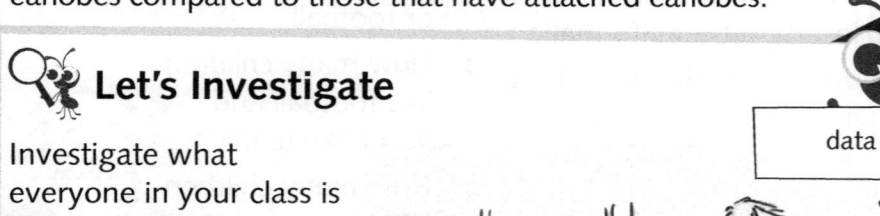

Investigate what everyone in your class is having for lunch.

Draw a table to show the most common types of food.

Write just one lunch menu that is representative of the most popular lunch for your class.

The Language of Maths

Look at this bar chart.
What might the data in the graph represent?
Label the vertical and horizontal axes and write a title for the graph.
Write an explanation about why your labels are appropriate for this set of data.

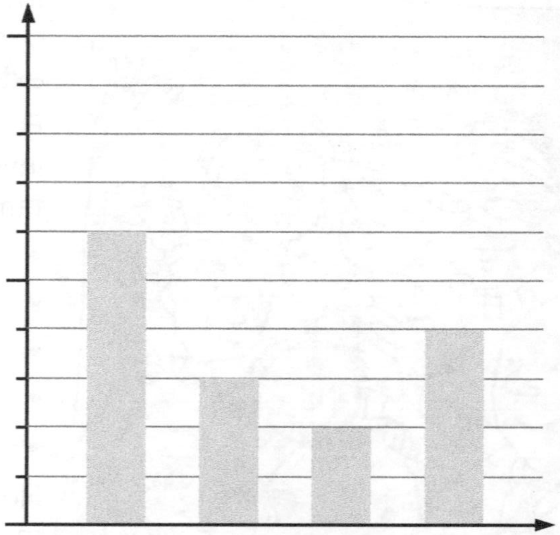

In the Past

Queen Elizabeth the Queen Mother, Queen Elizabeth II and Princess Anne all sat down for tea with the Prime Minister.

Draw a Carroll diagram to show their relationships to each other using the following terms:

Mother / Not mother
Daughter / Not daughter

Issue 35 – Statistics
S&C Volume 3

The Maths Herald

Name: Date:

The Language of Maths

The graph shows the number of sweets of each flavour in a packet of Fruit Flavoured Squares.

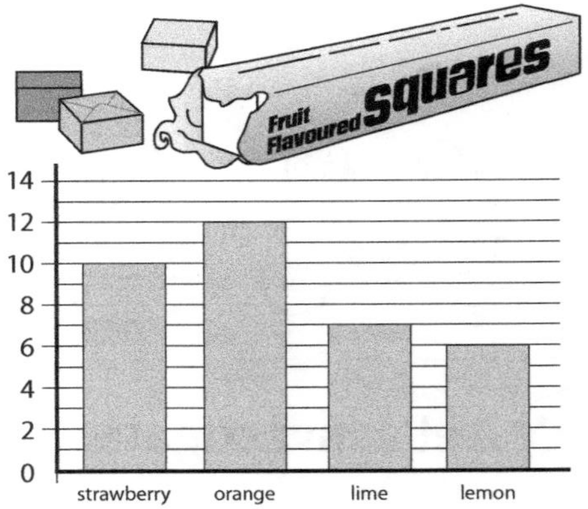

Using squared paper and 4 different coloured pencils, draw a diagram that shows the same information in a different way.

Write statements about the information shown on your diagram.

At Home

How long do you take to do things? Not running or solving problems, but things like eating your breakfast, brushing your teeth or getting ready for school in the morning?

Ask someone at home to time you as you do 4 different things, and write them in a table like the one on the right.

Activity	Time taken

© HarperCollins*Publishers* 2016

At Home

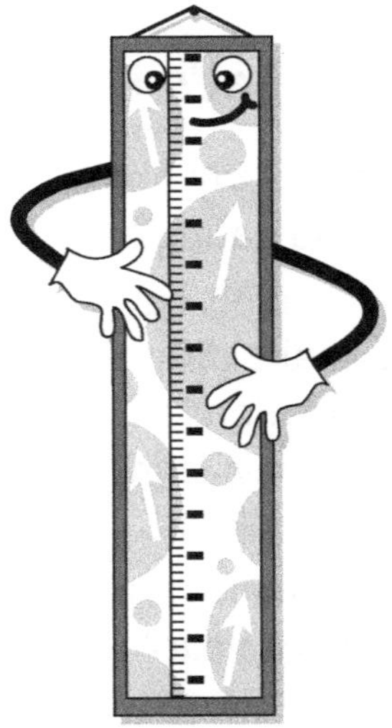

Draw a height chart to show how much you have grown each year.

To do this, you will need to talk to someone at home. Look at old photos, clothes or even an actual height chart.

Your graph will not be completely accurate, but try your best to draw your own height chart showing approximately how many centimetres you have grown each year.

The Arts Roundup

How do most people get a copy of the music they like to listen to?

Do they buy a CD, download it from the Internet, or use some other method?

Does this method differ for different age groups?

Write about how you are going to find out the answers to these questions.

Be prepared to justify your answers with evidence.

The Language of Maths

This Venn diagram shows what one class of children like to eat for breakfast. Write 8 statements about what the children like to eat.

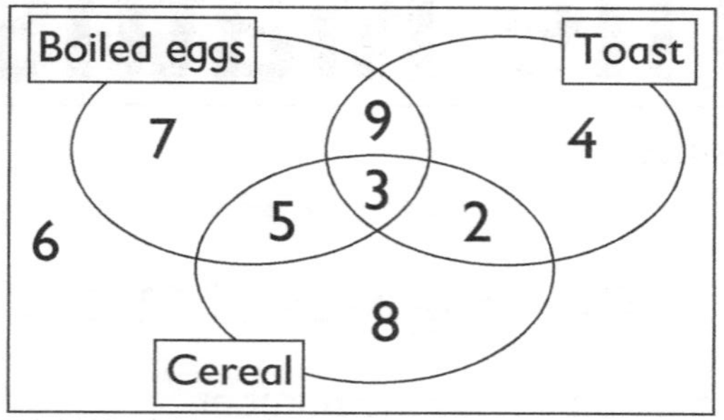

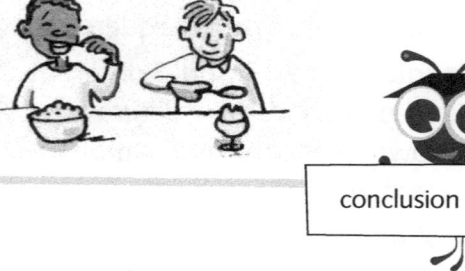

Boiled eggs   Cereal   Toast

7 9 5 3 2 4
   6   8

Let's Investigate

Do a survey of your class to see who likes apples, oranges and bananas. Some people may like only one of these fruit, some people may like two of them, some may like all 3 and some may like none of them.

Before you start, think about how you are going to record your information and how you are going to present your results.

conclusion

The Language of Maths

How many hours each week do you spend eating?

You might have to make different estimations for weekdays and the weekend.

How many hours is this a year?

Follow these steps to answer the question:

Step 1: Plan
Step 2: Collect data
Step 3: Organise the data
Step 4: Represent the data
Step 5: Interpret and discuss the data.

Tricky, tricky, tricky!

Let's Investigate

A, B, C, D, E, F, G, … W, X, Y, Z

Without the use of a dictionary, make a list of 26 mathematical words – one for each letter of the alphabet.

When you've made your list, compare your list to someone else's.

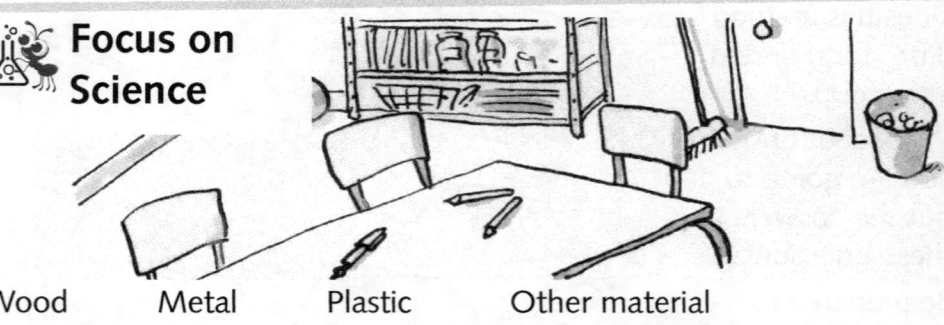

Focus on Science

Wood   Metal   Plastic   Other material

What are most of the objects in your classroom made from? Be prepared to justify your conclusions with evidence.

Issue 36 – Statistics
S&C Volume 3

The Maths Herald

Name: Date:

Around the World

Follow these steps to answer the question:

Step 1: Plan
Step 2: Collect data
Step 3: Organise the data
Step 4: Represent the data
Step 5: Interpret and discuss the data.

Write about what you did at each of the 5 steps and why you did it.

Where in the world would most children in your class like to go for a holiday?

Think about what destinations are likely to be the most popular.

pie chart

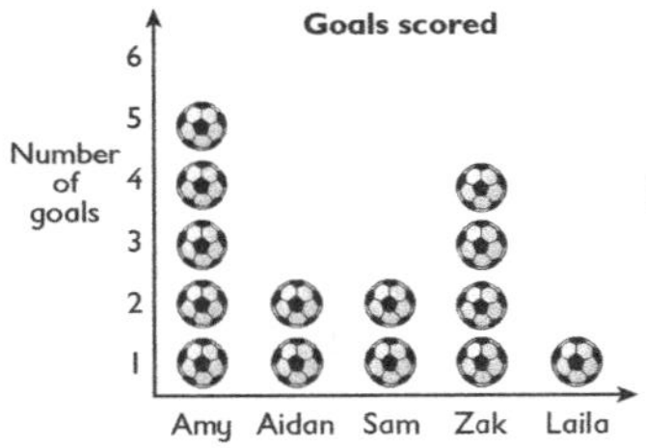

At Home

Make a complete list of all the electrical items in your home.

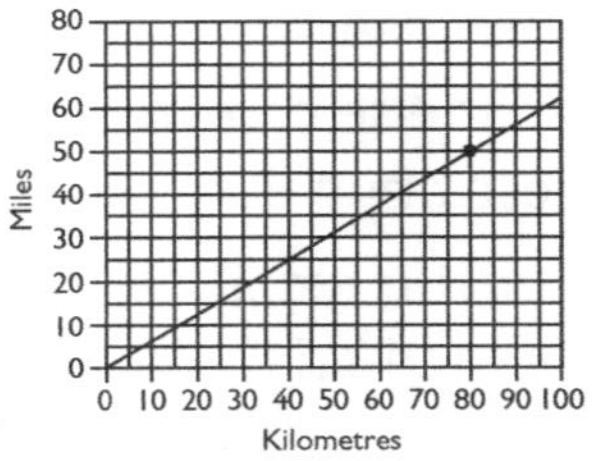

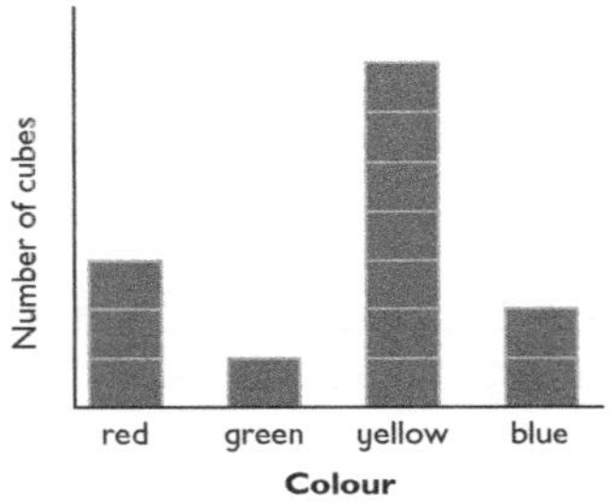

Once you have made a complete list, organise your list under different categories. What categories might you use?

Can you rewrite your list using different criteria?

© HarperCollins*Publishers* 2016

The Language of Maths

Look at these 12 tables, charts and graphs. What is each one called? Use the names at the bottom of the page to name each table, chart and graph.

Name	Age
Kiera	6 years 0 months
James	6 years 3 months
Jane	7 years 4 months
Sultan	6 years 11 months
Ashma	7 years 9 months
Asim	7 years 8 months

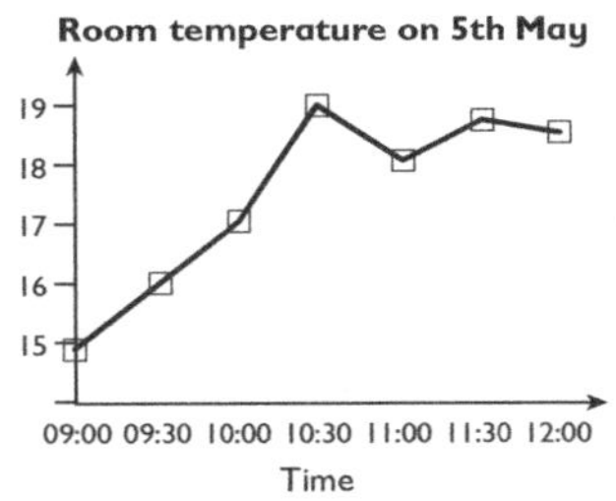

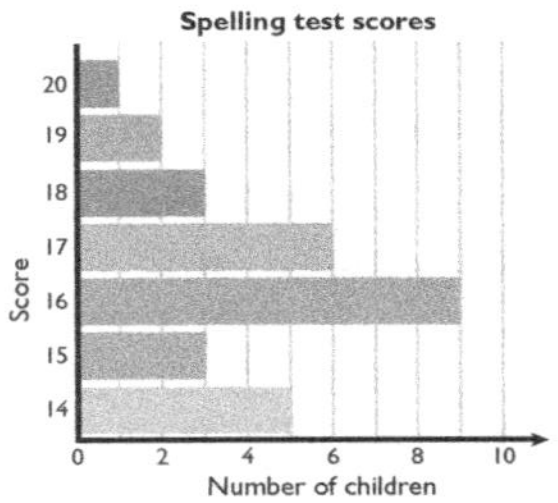

Class	Tally
Class A	ⅢⅠ ⅢⅠ
Class B	ⅠⅠⅠⅠ
Class C	ⅢⅠ ⅢⅠ ⅢⅠ ⅠⅠ
Class D	ⅠⅠⅠ

Room temperature on 5th May (line graph, Time vs temperature 09:00–12:00)

Spelling test scores (bar chart, Score 14–20, Number of children)

Kilometres to Miles conversion graph

Goals scored (pictogram: Amy, Aidan, Sam, Zak, Laila)

Venn diagram – Children aged 9 – Girls: Kala, Mila, Nina; Matt, Sheraz; Joel

Fruit	Number of children
apples	8
grapes	3
bananas	7
pears	6

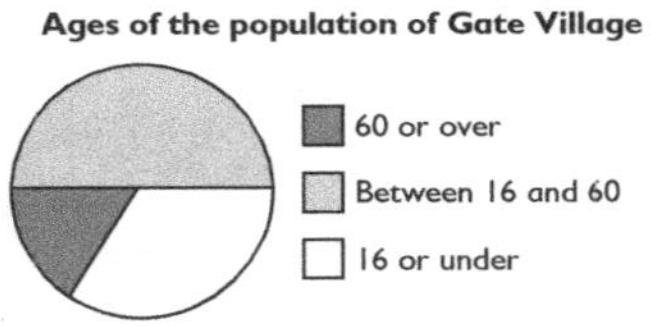

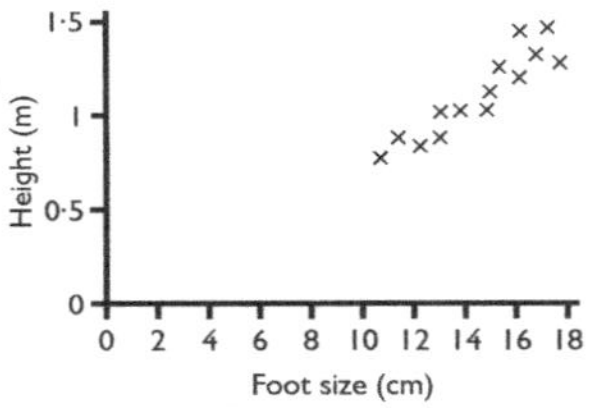

The fruit we like best

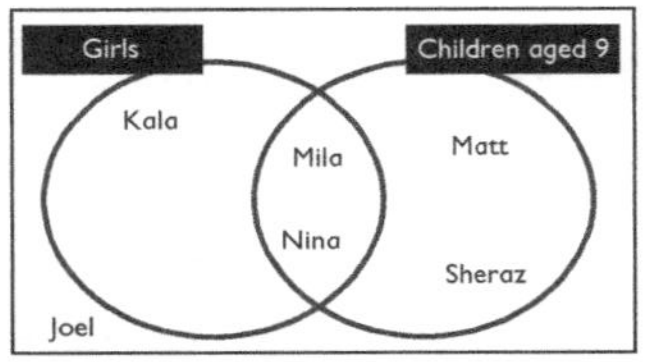

Scattergram: Foot size (cm) vs Height (m)

	Odd	Not odd (even)
Numbers that have 4 tens	47 41	48 40
Numbers that do not have 4 tens	23 25	26 20

Block graph: Number of cubes by Colour (red, green, yellow, blue)

Pie chart: Ages of the population of Gate Village (60 or over, Between 16 and 60, 16 or under)

- Bar chart
- Venn diagram
- Pie chart
- Carroll diagram
- Frequency table
- Scattergram
- Line graph
- Pictogram
- Block graph
- Conversion graph
- Table
- Tally chart

Issue 36 – Statistics

The Language of Maths

Big Bert and Large Les are 2 contestants in the Biggsmouth Hot Dog Competition.

The 2 charts below show the rate of hot dogs that each contestant ate during the competition.

Write different statements about Big Bert and Large Les at the Biggsmouth Hot Dog Competition.

Big Bert — Number of hot dogs eaten each minute vs Time

Large Les — Number of hot dogs eaten each minute vs Time

50th Annual Biggsmouth Hot Dog Competition

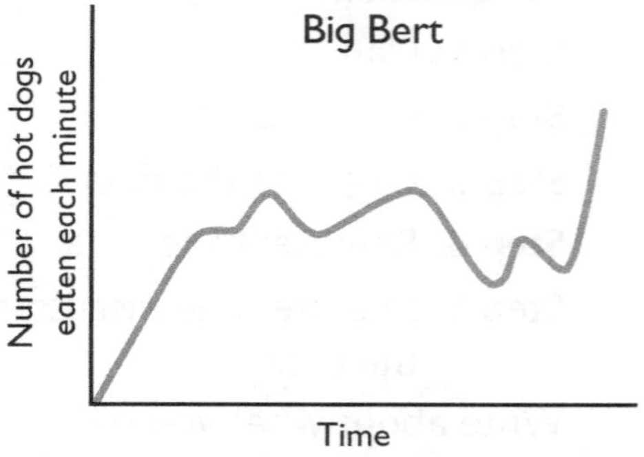

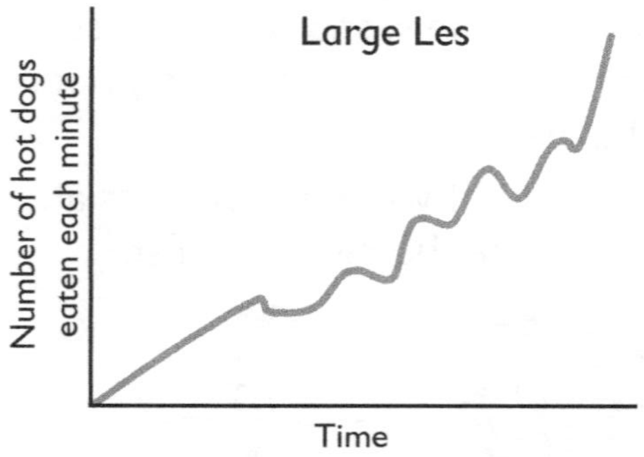

Focus on Science

Draw and label a Venn diagram similar to the one on the right.

Write the names of all the different types of fruit and vegetables and other foods that you can think of on the diagram.

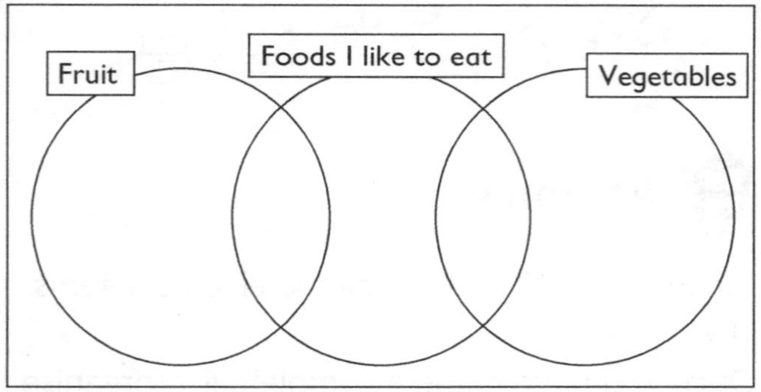

Venn diagram: Fruit | Foods I like to eat | Vegetables

Focus on Science

Animals can be classified in different ways.

One method of classification is shown on the right.

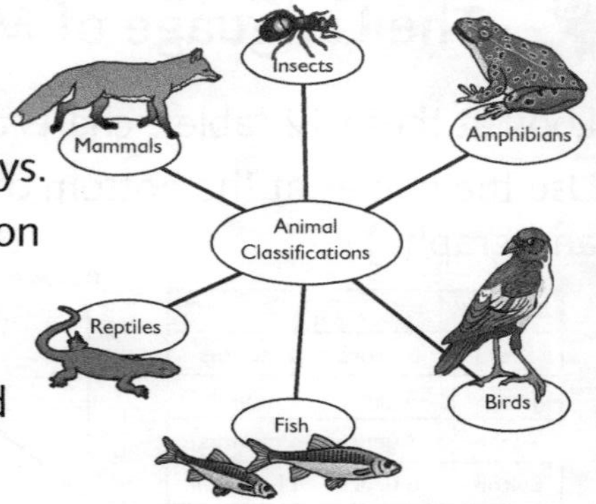

Investigate different animals. Find out to which animal group they belong and whether they live on land or on water.

Then draw a Carroll diagram similar to the one on the right to record your results.

What else can you find out about these animals?

Can you record your results in a different Carroll diagram using other criteria? What about in a Venn diagram?

Animal Classifications: Mammals, Insects, Amphibians, Birds, Fish, Reptiles

Carroll diagram:
	Mammal	Not a mammal
Lives on land		
Does not live on land		

Let's Investigate

An activity to do with a friend.

Take these 4 cards from a pack of playing cards and lay them out face up on the table.

Use the cards to answer these questions.

- If you turned over just 1 card:
  - how likely is it that the card would be the King of Hearts?
  - how likely is it that the card would be a black card?
  - are you more likely to choose the Ace of Spades or a red card?

Write an explanation for your answers above.

Now shuffle the cards and lay them out face down on the table.

Imagine that the 4 cards were shuffled and laid out face down on the table.

Investigate how accurate your predictions to the above questions were. Write about your results.

© HarperCollins*Publishers* 2016

Teacher's notes

Issue 1

Number

Prerequisites for learning

- Identify patterns and relationships involving numbers
- Read, write and order whole numbers to at least 1000
- Count on from and back to zero in single-digit steps or multiples of 10
- Recognise the place value of each digit in a three-digit number (hundreds, tens, ones)
- Estimate a number of objects
- Recognise multiples

Resources

pencil and paper
Resource sheet 2: My notes (optional)
Resource sheet 3: Pupil self assessment booklet (optional)
14 counters, or similar (optional)

Teaching support

Page 1

Let's Investigate

- Provide the children with 14 counters (or similar).
- Remind the children that the maximum number of beads that they can have in each column of the abacus is 9.

The Puzzler

- Tell the children the correct position of some of the digits (see Answers).

Page 2

Looking for Patterns

- Once children have completed the activity, discuss the various methods they used to determine a particular shape in the pattern, and also the position of a given shape in the pattern.
- Give the children different patterns involving squares and triangles, and ask them to answer the same set of questions, for example:

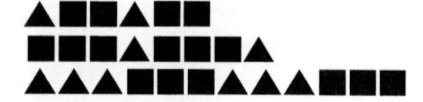

▲■■▲■■
■■■▲■■■▲
▲▲▲▲■■■▲▲▲■■■

Looking for Patterns

- Encourage the children to make patterns involving three or more different criteria, for example:

◆✳✳◉◯◆✳✳◉◯
▲◉✳✳◆◇◉▲◉✳✳◆◇◉▲

Looking for Patterns

- Children use only one rule for each number sequence, e.g. 1, 5, 9, 13 (rule: + 4).
- Children use two rules for each number sequence, e.g. 1, 3, 7, 15 (rule: × 2, + 1).

Page 3

The Puzzler

- Encourage the children to make their own puzzle using as few clues as possible.

96

Let's Investigate

- Children investigate other objects in the school, for example, the total number of chairs, tables, paintbrushes, interlocking cubes …

Looking for Patterns

- Suggest the children look at the difference between consecutive numbers in the sequence. What pattern do they notice? (Each consecutive difference is 50 more than the previous difference.)

    30,   80,   180,   330, …
         50   100   150

Page 4

What's the Problem?

- Suggest the children make a systematic list to represent the total number of handshakes, for example:

Mr Keft – Mr A	Mrs Keft – Mr A	Mr A – Mr B	Mrs A – Mr B
Mr Keft – Mrs A	Mrs Keft – Mrs A	Mr A – Mrs B	Mrs A – Mrs B
Mr Keft – Mr B	Mrs Keft – Mr B	Mr A – Mr C	Mrs A – Mr C
Mr Keft – Mrs B	Mrs Keft – Mrs B	Mr A – Mrs C	Mrs A – Mrs C
Mr Keft – Mr C	Mrs Keft – Mr C		
Mr Keft – Mrs C	Mrs Keft – Mrs C		
Mr B – Mr C	Mrs B – Mr C		
Mr B – Mrs C	Mrs B – Mrs C		

Money Matters

- Most children should find this problem relatively easy. Once they have solved it, discuss with them their strategies for finding the amounts.

What's the Problem?

- Tell the children to read through all the instructions carefully. Discuss with them how using the clues in a different order may help them arrive at the number more easily and quickly.
- Tell the children that there are actually two numbers inside the box. Can they work out what both numbers are?
- Children make up a similar puzzle of their own and give it to a friend to solve. However, tell the children that their clues must lead to one number only (not two or more). Can they make such a problem using the fewest number of clues?

AfL

- What was the largest / smallest number you could make? How do you know it's the largest / smallest?
- What is happening in this sequence?
- What pattern do you notice? What is the rule?
- How did you work out what the shape was?
- What are all the different possibilities? How can you be sure that you have accounted for them all?
- Which of the clues really helped you to narrow down what the number might be?
- How did you arrive at your approximation / estimation?

Issue 1 – Number

Answers

Page 1

Let's Investigate
The largest number you can make on the abacus is 950.
The smallest number you can make on the abacus is 59.

The Puzzler

			42 ▼	268 ▼		79 ▼	332 ▼	
652 ►	5	2	6	394 ►	4	9	3	816 ▼
	2	984 ►	4	8	9	217 ►	7	2	1
95 ►	5	9	123 ►	2	3	1	63 ►	3	6
436 ►	6	4	3	249 ►	4	2	9	421 ►	8
	928 ►	8	2	9	175 ►	7	5	1	8359 ▼
	2	415 ►	1	4	5	485 ►	8	4	5
58 ►	8	5	125►	2	1	5	82 ►	2	8
169►	9	1	6	827 ►	7	8	2		9
	04 ►	4	0		2483►	4	8	2	3

Page 2

Looking for Patterns
The 40th shape is a square.
The 46th shape is a triangle.
The position of the 15th square is the 25th shape.
A triangle is to the right of the 15th square.
The 83rd shape is a square.
The position of the 31st triangle is the 76th shape.

Looking for Patterns
Printer's wheel puzzles and questions will vary.

Looking for Patterns
The rule is: + 4. The next three numbers are: 17, 21, 25.
Number sequences will vary.

Page 3

The Puzzler
The number is 224.
Number puzzles will vary.

Let's Investigate
Results of the investigation will vary.

Looking for Patterns
The rule is each consecutive difference is 50 more than the previous difference.
The next four numbers in the sequence are 530, 780, 1080, 1430.

Page 4

What's the Problem?
There are 24 handshakes altogether.

Money Matters
Luis has £21 pocket money and Marcus has £7 pocket money.

What's the Problem?
There are actually two numbers in the box: 634 and 652.

Inquisitive ant

number sequence
A pattern of numbers that follow on one after the other.

Issue 2
Number

Prerequisites for learning

- Identify patterns and relationships involving numbers
- Read, write and order whole numbers to at least 1000
- Count on from and back to zero in single-digit steps or multiples of 10
- Recognise the place value of each digit in a three-digit number (hundreds, tens, ones)
- Use <, > and = signs
- Estimate a number of objects

Resources

pencil and paper
Resource sheet 2: My notes (optional)
Resource sheet 3: Pupil self assessment booklet (optional)
counters, or similar (optional)

Teaching support

Page 1

The Puzzler

- Do not encourage the children to find the solution to this puzzle by counting on from 20 until they reach 76 to find out which box the ticket will fall from. Instead, suggest they look for patterns and use mental calculation strategies to find the solution. Do they notice that the numbers in each box increase by 6 each time? Can they use this to work out from which box number 76 will fall?

Looking for Patterns

- Assist children in visualising the problem by drawing a simple diagram like this.
- An important aspect of this activity is the children's ability to write an explanation of how they worked out the answer to the problem.

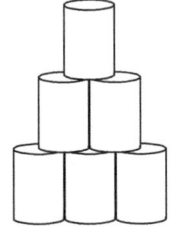

Page 2

The Puzzler

- Ensure children understand how to read the less than and greater than signs in the bottom right-hand corner of the grid of squares as shown.

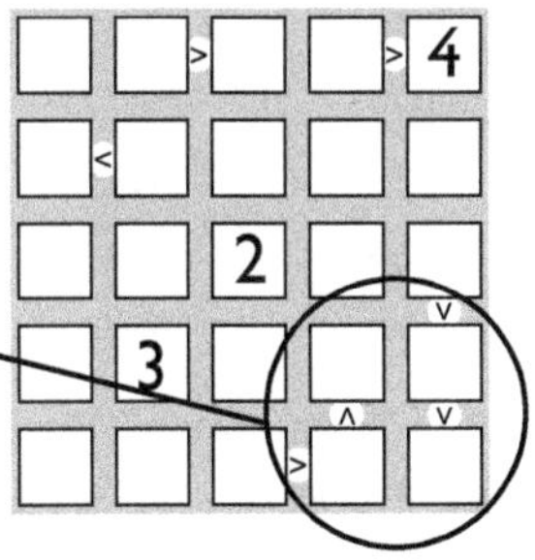

Looking for Patterns

- One method of identifying the pattern is to look at the difference between rows and columns, that is:

1	2	3	4	5	6	7	8	9	10
36	37	38	39	40	41	42	43	44	11
35	64	65	66	67	68	69	70	45	12
34	63						71	46	13
33	62						72	47	14
32	61		?				73	48	15
31	60							49	16
30	59							50	17
29	58	57	56	55	54	53	52	51	18
28	27	26	25	24	23	22	21	20	19

The difference between most of the pairs of numbers in the 1st and 2nd columns is 29.

The difference between most of the pairs of numbers in the 1st and 2nd rows is 35.

The difference between most of the pairs of numbers in the 9th and 10th columns is 33.

The difference between most of the pairs of numbers in the 9th and 10th rows is 31.

99

Issue 2 – Number

- The pattern continues as shown:

				↕ Difference of 35		
				↕ Difference of 27		
				↕ Difference of 19		
	↕ Diff of 29	↕ Diff of 21	↕ Diff of 13			↕ Diff of 17	↕ Diff of 25	↕ Diff of 33
				↕ Difference of 15		
				↕ Difference of 23		
				↕ Difference of 31		

- Once children have identified a pattern, arrange them into pairs to share their ideas and discuss their reasoning.

Page 3

Looking for Patterns

- 22 guestroom doors have the number 7 on them:
  107, 207, 307, 407, 507, 607, 701, 702, 703, 704, 705, 706, 707, 708, 709, 710, 711, 712, 713, 714, 715, 807.

80 guestroom doors have a zero on them:

101	201	301	401	501	601	701	801
102	202	302	402	502	602	702	802
103	203	303	403	503	603	703	803
104	204	304	404	504	604	704	804
105	205	305	405	505	605	705	805
106	206	306	406	506	606	706	806
107	207	307	407	507	607	707	807
108	208	308	408	508	608	708	808
109	209	309	409	509	609	709	809
110	210	310	410	510	610	710	810

The Language of Maths

- Ensure children are familiar with the term "digit" and know that there are 10 digits (0–9).
- Ask the children to make up similar questions for a friend to solve.

Page 4

Looking for Patterns

- Suggest children draw a diagram.
- How many elves are wearing both pointy shoes and green caps? (1)

Let's Investigate

- Discuss with the children how they might keep a record of the number of buttons they count. If appropriate, suggest that the children draw a tally chart.
- Remind the children that they are only expected to find the approximate number of buttons, i.e. an estimate, but that their estimate should be as accurate as possible.
- When pairs of children have arrived at their estimate, as a group, ask different pairs to discuss what they did. Which estimates are probably the most accurate? Why?

Sports Update

- What if the final score was 1–1?
- Children can use counters (or similar) to represent the half-time scores.
- Once children recognise that the total number of possible half-time scores for both matches are square numbers, i.e. 9 and 25, ask the children to work out a rule for finding the total number of possible half-time scores for any match that ends in a draw (one more than the draw score squared, e.g. final score of 3–3: $4^2 = 16$ different half-time scores possible; final score of 5–5: $6^2 = 36$ different half-time scores possible). Can the children apply this rule to find out the total number of possible half-time scores for a match with a final score of 10–10? (121)

AfL

- What patterns do you notice? How does this help you?
- How could you show the solution to this problem using a diagram?
- How did you work out the answer to this problem? What patterns did you recognise? What generalisations were you able to make?
- How did you arrive at your approximation / estimation?
- What are all the different possibilities? How do you know that you have accounted for them all?
- Were you systematic in the way you worked out the answer to this problem? What did you do to help you organise your work?

Answers

Page 1

The Puzzler
Ticket number 76 will fall from box D.

Looking for Patterns
If there are 36 cans on display, then there are 8 rows of cans.
If there were 12 rows of cans, there would be 78 cans on display.

Page 2

The Puzzler

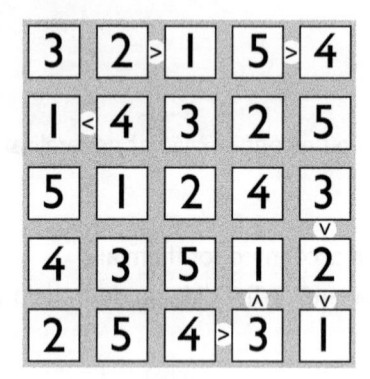

3	2 >	1	5 >	4
1 <	4	3	2	5
5	1	2	4	3
4	3	5	1	2
2	5	4 >	3	1

Looking for Patterns
? = 95

Page 3

Looking for Patterns
There are 22 guestroom doors that have the number 7 on them.
80 guestrooms of the hotel have a zero on them.

The Language of Maths
9992 is the largest four-digit number that has 2 as one of its digits.
1002 is the smallest four-digit number that has 2 as one of its digits.
9876 is the largest four-digit number in which no digit occurs more than once.
1023 is the smallest four-digit number in which no digit occurs more than once.
9872 is the largest four-digit number that has a 2 and in which no digit occurs more than once.
1023 is the smallest four-digit number that has a 2 and in which no digit occurs more than once.

Page 4

Looking for Patterns
Six elves are not wearing either pointy shoes or green caps.

Let's Investigate
Results of the investigation will vary.

Sports Update
There are nine possible half-time scores with a final score of 2–2:
0–0, 0–1, 1–0, 0–2, 2–0, 1–1, 1–2, 2–1, 2–2.
There are 25 possible half-time scores with a final score of 4–4:
0–0, 0–1, 0–2, 0–3, 0–4,
1–0, 2–0, 3–0, 4–0,
1–1, 1–2, 1–3, 1–4,
2–1, 2–2, 2–3, 2–4
3–1, 3–2, 3–3, 3–4
4–1, 4–2, 4–3, 4–4.
Both the numbers 9 and 25 are square numbers.

Inquisitive ant 
≤ Less than or equal to.

Issue 3
Number

Prerequisites for learning

- Identify patterns and relationships involving numbers
- Read, write and order whole numbers to at least 1000
- Use <, > and = signs
- Estimate a number of objects
- Begin to understand the terms: "mixed number", "square number" and "multiple"
- Solve logic puzzles

Resources

pencil and paper
Resource sheet 2: My notes (optional)
Resource sheet 3: Pupil self assessment booklet (optional)
selection of different take-away menus
art paper and colouring materials
computer with Internet access

Teaching support

Page 1

Let's Investigate

- Ensure children complete this activity before starting on the next activity (Construct).
- Children need to be provided with a wide selection of different take-away menus to compare so that they can accurately evaluate which menus offer good value for money and which menus are expensive.
- The important aspect of this activity is the justifications children offer for their conclusions.

Construct

- Ensure children have completed the Let's Investigate activity before starting on this activity.
- Children can create their menu using ICT.
- When pairs of children have completed their menu they can compare their menus with another pair's. Are the prices realistic? Do the menus offer good value for money? Can the children say why / why not?

Page 2

Looking for Patterns

- There are a total of 32 tickets with the number 3 on them:
    Ticket numbers: 3, 13, 23 (3)
    Ticket numbers: 30–39 (10)
    Ticket numbers: 43, 53, 63, 73, 83, 93, 103, 113, 123 (9)
    Ticket numbers: 130–139 (10)
    Total: 3 + 10 + 9 + 10 = 32
- The number 3 appears a total of 34 times on the 140 tickets – the 32 times mentioned on the tickets above, plus an additional 3 on ticket numbers 33 and 133.

Let's Investigate

- Remind the children that they are only expected to find an approximation.
- Ask the children how they could make their approximations more accurate.

Sports Update

- This activity introduces children to the language of ratio and proportion. If appropriate, you may wish to take this concept further by asking questions such as: "What if there were four women for every one man?"

Issue 3 – Number

- Suggest the children organise their working out using a diagram or in a table, for example:

	www m	www m	www m	www m	www m
Women	3	6	9	12	15
Men	1	2	3	4	5
Total	4	8	12	16	20

Page 3

The Puzzler

- Ensure children are familiar with the less than (<) and greater than (>) signs.
- Children make up similar problems for a friend to solve. However, unlike the problems in the issue, children provide the inequalities rather than the numbers, for example:

☐ < ☐ > ☐ > ☐ < ☐   or   ☐ > ☐ > ☐ < ☐ < ☐

Around the World

- Before setting the children to work independently on this activity, ensure that they understand the concept of negative numbers and the terms "minimum temperature", "maximum temperature" and "Celsius", and if appropriate "Fahrenheit".
- You may wish to suggest to the children that they record their findings in a table, for example:

City / Town	Maximum temperature	Minimum temperature

- Once children have completed the activity, discuss with them the reasons for the differences in order between the two lists.

Page 4

The Language of Maths

- Ensure children have some understanding of most of the statements. However, part of the challenge of this activity is for the children to match the number with the statement through a process of elimination.
- Children create their own lists of numbers and statements. Encourage them to think of a range of different types of appropriate mathematical statements.

The Arts Roundup

- Ask the children to write a similar problem, e.g. the number of children in six different classes; the number of tickets sold to five different films; the number of people in five queues at a supermarket …

AfL

- What did you find out about menus? Which meals are expensive / cheap / offer good value for money?
- How does your menu compare? Are your prices reasonable? Why do you say that?
- Are you sure that you have found all the possibilities? Why are you so sure?
- How did you arrive at your estimation / approximation? How accurate do you think it is?
- Which of your cities has the coldest / warmest minimum / maximum temperature?
- Which cities are colder / warmer than …?

Answers

Page 1

Let's Investigate
Results of the investigation will vary.

Construct
Menus will vary.

Page 2

Looking for Patterns
There are 32 tickets with the number 3 on them.
The number 3 appears 34 times.

Let's Investigate
Results of the investigation will vary.

Sports Update
There were 15 women and 5 men in the class.

Page 3

The Puzzler

9 > 4 < 7 > 3 < 5

17 > 2 > 1 < 8 > 7

7 < 11 > 10 > 8 < 12

11 < 12 > 4 > 3 < 6

2 < 3 < 14 > 6 < 7

8 > 7 > 6 < 10 < 14

Around the World
Results of the investigation will vary.

Page 4

The Language of Maths
The majority of numbers fit more than one statement.
Answers will vary.

The Arts Roundup
The Palace Theatre – 635
The Actor's Space – 314
The Playhouse – 286
The Queen's Centre – 528
The Studio – 157

Inquisitive ant

estimate
A rough calculation or judgement about an answer, value, number or quantity.

Issue 4
Number

Prerequisites for learning

- Identify patterns and relationships involving numbers
- Read, write and order whole numbers to at least 1000
- Recognise the place value of each digit in a three-digit number (hundreds, tens, ones)
- Solve logic puzzles

Resources

pencil and paper
Resource sheet 2: My notes (optional)
Resource sheet 3: Pupil self assessment booklet (optional)
selection of different newspapers and magazines

Teaching support

Page 1

What's the Problem?

- There are three numbers less than 4 (1, 2, 3) and five numbers greater than 4 (5, 6, 7, 8, 9) so the product is 5 × 3 = 15, the same as it would be if the competitor's number was 6, with five numbers less and three numbers greater.

Technology Today

- Is the correlation between the digits and letters on each key the same for all phones? Why do you think this is?

Page 2

Famous Mathematicians

- Ensure children can recognise triangular and square numbers.

What's the Problem?

- This activity introduces children to the language of ratio and proportion. If appropriate, you may wish to take this concept further by discussing the idea of using sequences to scale numbers up or down.
- Suggest the children organise their working out in a table, for example:

Bulbs	6	12	18	24	30				
Plants	5	10	15	20	25				

- Are the children able to identify a pattern and apply this to work out the answer? For example:

  6 × ☐ = 120

  Therefore ☐ = 20

  So 5 × 20 = 100

Page 3

Let's Investigate

- Encourage the children to look for examples of decimals in a variety of contexts.
- The financial section of a newspaper offers a range of decimals to one, two, three and more decimal places.
- The most important aspect of this activity is in the children offering explanations as to the meaning of each of the decimals. Some of these will be self-evident, such as in the context of money; however, others will not be so obvious. As a result, allow the children to work in pairs so that they can share ideas and discuss their reasoning.

The Puzzler
- When writing their own puzzles, encourage the children to be as imaginative as they can when choosing what criteria to use for determining the digit in each place value in the number.

Focus on Science
- Suggest children use trial and improvement to work out the number of each beetle and spider, for example:

5 spiders and 5 beetles

$= (5 \times 8) + (5 \times 6)$

$= 40 + 30$

$= 70$

Not enough legs, so:

$= (6 \times 8) + (4 \times 6)$

$= 48 + 24$

$= 72$

Still not enough legs, so …

Page 4

Let's Investigate
- Tell the children that including 123, there are a total of 21 three-digit numbers whose digits total 6.
- Encourage the children to be systematic when investigating all the possible numbers.

The Language of Maths
- Although this problem involves no actual mathematics, it does require the children to problem solve, reason and think logically.
- The first person's statement can't be true as they say that both of them are Nays and therefore always lie (if they were both Nays, they would lie about it). So the first person is a Nay and their statement is a lie. Therefore the second person must be a Yay.
- Point out that the two people may belong to different tribes.

The Language of Maths
- Tell the children that for both sequences each of the letters is the first letter of a word.
- Children make up similar sequences of their own for a friend to solve.

AfL

- Tell me one of your phone number codes and let's see if I can decode it.
- How did you work out the answer / solution to this problem / puzzle?
- What can you tell me about triangular numbers?
- What patterns did you notice about the numbers?
- How did you work out that this digit was a 6? Tell me some of the clues that you wrote for a number.
- Have you found all the numbers possible? How do you know that you have? How did you organise your work as you did this investigation?
- What are the rules for this sequence?

Answers

Page 1

What's the Problem?
The competitor's number is 4.

Technology Today

1 oo	2 abc	3 def
4 ghi	5 jkl	6 mno
7 pqrs	8 tuv	9 wxyz
* +	0 ␣	# ⇧

season: summer
day: Saturday
month: November
planet: Neptune
fruit: apricot
vegetable: potato
city: Liverpool
country: Thailand

The children's codes will vary.

Page 2

Famous Mathematicians
When you multiply any triangular number by 8 and add 1, the result is always a square number.

What's the Problem?
100 out of the 120 bulbs grew into tulips.

Page 3

Let's Investigate
Results of the investigation will vary.

The Puzzler
6875   7385

Focus on Science
There are seven spiders and three beetles.

Page 4

Let's Investigate

105	114	123	204	222	303	600
150	141	132	240		330	
501	411	213	402			
510		231	420			
		312				
		321				

Explanations will vary.

The Language of Maths
The first is a Nay and the second is a Yay.

The Language of Maths
F, M, A, M, J, J, A, S, O, N (first letter of the months of the year, in order, starting with February)
o, t, t, f, f, s, s, e, n, t (first letter of the counting sequence of numbers, starting with one)

Inquisitive ant

decimal
A fraction with ten as its denominator. A decimal point separates the whole numbers from the decimal fraction, e.g. $\frac{1}{2} = \frac{5}{10} = 0.5$.

Issue 5
Addition

Prerequisites for learning

- Identify patterns and relationships involving numbers
- Recall and use addition and subtraction facts to 20 fluently, and derive and use related facts up to 100
- Add and subtract numbers mentally, including:
  - a two- or three-digit number and ones
  - a two- or three-digit number and tens
  - two two-digit numbers
  - adding three one-digit numbers
  - a three-digit number and hundreds
- Add and subtract numbers with up to three digits, using formal written methods of columnar addition and subtraction

Resources

pencil and paper
Resource sheet 2: My notes (optional)
Resource sheet 3: Pupil self assessment booklet (optional)
Resource sheet 9: 2 cm squared paper (optional)
ruler (optional)
set of dominoes

Teaching support

Page 1

Let's Investigate

- Do the children realise that the number in the top brick is always the sum of the number in the first bottom brick, plus twice the number in the middle bottom brick, plus the number in the third bottom brick?
- Choose three numbers from the bricks and work through an example with the children, for example:

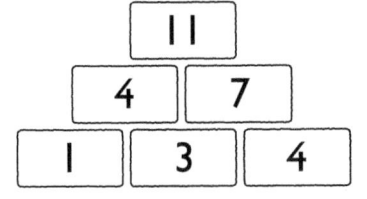

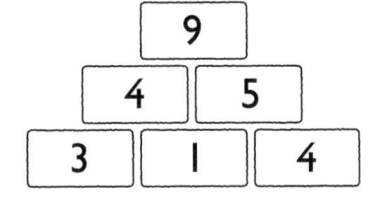

- Children try product or difference walls rather than addition walls.
- Children try walls of different sizes, for example:

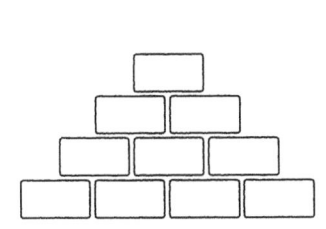

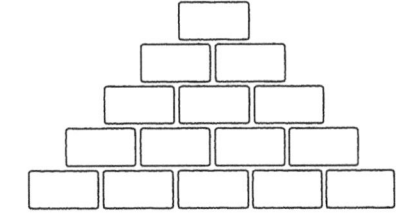

The Puzzler

- Once individual children have solved the puzzle, discuss as a group the strategies that different children used. Which strategies were most efficient?
- Children create a similar puzzle for a friend to solve.

Page 2

Let's Investigate

- Ensure children realise that this activity requires them to include all the addition number facts for 0, 1, 2, 3, 4, 5, 6, 7, 8, 9 and 10.

109

Issue 5 — Addition

- Discuss with the children the patterns they notice in the addition number facts for 5. This will help them to identify the addition facts for other numbers.
- Although all the addition number facts for 5 are included in the issue as an example, if necessary work with the children to identify all the addition facts for another number, e.g. 8 (see Answers).
- Tell the children that altogether there are 66 addition number facts to 10.

Let's Investigate

- You may wish to provide the children with 2 cm squared paper on which to record their calculations.
- Children make subtraction and multiplication calculations, for example:

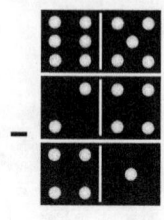

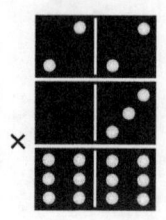

Looking for Patterns

- Ensure children are familiar with the term "consecutive numbers", and in particular the idea of consecutive even numbers and consecutive odd numbers.
- If children have difficulty in identifying the patterns, tell them to look at the units digits in all the answers in both sets of calculations. Do they realise that for the calculations involving three consecutive odd numbers, the totals all have an odd units digit? Similarly, do they notice that for the calculations involving three consecutive even numbers, the totals all have an even units digit?

Page 3

The Puzzler

- Tell the children that each of the numbers in the six grey boxes is less than 10.

The Puzzler

- Tell the children one of the numbers that both Fabio and Gabby hit (see Answers).

Page 4

Let's Investigate

- After children have been working on the investigation for some time, discuss with them what they notice. Do they realise that they do not need to investigate all two-digit numbers – that if a number such as 23 is a one-stage palindromic number, then the number 32 is also a one-stage palindromic number?
- Through a process of elimination, the children should realise that it is the numbers 89 and 98 that become palindromic after 24 stages.

AfL

- What different totals were you able to make? What was the largest total / smallest total you made? Could you make an even larger / smaller total?
- How did you work out this puzzle? How did you start?
- What are all the addition number facts for 6? What do you notice about these facts? What relationships can you see?
- What is similar about these two facts?
- What different calculations were you able to make?
- What patterns did you notice? How did these help you?

Issue 5 – Addition

Answers

Page 1

Let's Investigate
Results of the investigation will vary.

The Puzzler

1	5	3	(9)
7	2	8	(17)
4	6	9	(19)

(12) (13) (20)

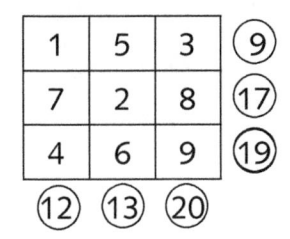

Page 2

Let's Investigate
There are 66 addition number facts up to, and including, 10.

0 + 0 = 0	1 + 0 = 1	2 + 0 = 2	3 + 0 = 3
	0 + 1 = 1	1 + 1 = 2	2 + 1 = 3
		0 + 2 = 2	1 + 2 = 3
			0 + 3 = 3

4 + 0 = 4	5 + 0 = 5	6 + 0 = 6	7 + 0 = 7
3 + 1 = 4	4 + 1 = 5	5 + 1 = 6	6 + 1 = 7
2 + 2 = 4	3 + 2 = 5	4 + 2 = 6	5 + 2 = 7
1 + 3 = 4	2 + 3 = 5	3 + 3 = 6	4 + 3 = 7
0 + 4 = 4	1 + 4 = 5	2 + 4 = 6	3 + 4 = 7
	0 + 5 = 5	1 + 5 = 6	2 + 5 = 7
		0 + 6 = 6	1 + 6 = 7
			0 + 7 = 7

8 + 0 = 8	9 + 0 = 9	10 + 0 = 10
7 + 1 = 8	8 + 1 = 9	9 + 1 = 10
6 + 2 = 8	7 + 2 = 9	8 + 2 = 10
5 + 3 = 8	6 + 3 = 9	7 + 3 = 10
4 + 4 = 8	5 + 4 = 9	6 + 4 = 10
3 + 5 = 8	4 + 5 = 9	5 + 5 = 10
2 + 6 = 8	3 + 6 = 9	4 + 6 = 10
1 + 7 = 8	2 + 7 = 9	3 + 7 = 10
0 + 8 = 8	1 + 8 = 9	2 + 8 = 10
	0 + 9 = 9	1 + 9 = 10
		0 + 10 = 10

Let's Investigate
Results of the investigation will vary.

Looking for Patterns

1 + 3 + 5 = 9	2 + 4 + 6 = 12
3 + 5 + 7 = 15	4 + 6 + 8 = 18
5 + 7 + 9 = 21	6 + 8 + 10 = 24
7 + 9 + 11 = 27	8 + 10 + 12 = 30
23 + 25 + 27 = 75	32 + 34 + 36 = 102

Explanations will vary.

Page 3

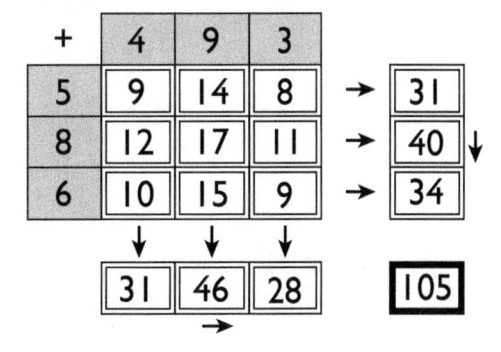

The Puzzler

+	4	9	3	
5	9	14	8	→ 31
8	12	17	11	→ 40
6	10	15	9	→ 34

↓ ↓ ↓

| 31 | 46 | 28 | | 105 |

The Puzzler
Fabio hit the numbers 57, 32, 41 and 16.
Gabby hit the numbers 78, 63, 29 and 16.

Page 4

Let's Investigate
One-stage palindromic numbers: 10, 12, 13, 14, 15, 16, 17, 18, 20, 21, 23, 24, 25, 26, 27, 29, 30, 31, 32, 34, 35, 36, 38, 40, 41, 42, 43, 45, 47, 50, 51, 52, 53, 54, 56, 60, 61, 62, 63, 65, 70, 71, 72, 74, 80, 81, 83, 90 and 92.
Two-stage palindromic numbers: 19, 28, 37, 39, 46, 48, 49, 57, 58, 64, 67, 73, 75, 76, 82, 84, 85, 91, 93, 94.
Three-stage palindromic numbers: 59, 68, 86, 95
Four-stage palindromic numbers: 69, 78, 87 and 96.
Five-stage palindromic numbers: There are no two-digit numbers.
Six-stage palindromic numbers: 79 and 97.
The numbers 89 and 98 only become palindromic after 24 stages.

Inquisitive ant

commutative law
For addition and multiplication, the same result occurs regardless of the order in which the numbers occur, e.g. 2 + 3 = 3 + 2 and 4 × 5 = 5 × 4.

Issue 6
Addition

Prerequisites for learning

- Identify patterns and relationships involving numbers
- Recall and use addition and subtraction facts to 20 fluently, and derive and use related facts up to 100
- Add and subtract numbers mentally, including:
  - a two- or three-digit number and ones
  - a two- or three-digit number and tens
  - two two-digit numbers
  - adding three one-digit numbers
  - a three-digit number and hundreds
- Add and subtract numbers with up to three digits, using formal written methods of columnar addition and subtraction

Resources

pencil and paper
Resource sheet 2: My notes (optional)
Resource sheet 3: Pupil self assessment booklet (optional)
Resource sheet 9: 2 cm squared paper (optional)
ruler
set of dominoes (optional)
calculator (optional)

Teaching support

Page 1

The Puzzler

- AAH must be a number greater than 499 as its double is a four-digit number.
  HARP must be a number between 1000 and 1999 as it is a four-digit sum of two identical three-digit numbers.
  So H = 1 and A = 5, 6, 7, 8 or 9. Insert these numbers into the calculation and only 991 fits.
- Tell the children the value of one of the letters (see Answers).
- Allow the children to use a calculator.

Money Matters

- Tell the children to show the denominations of the coins by writing them inside the circles, for example:

  1p   5p   20p

Page 2

Money Matters

- Children spend less than £5, e.g. £3.
- Children spend more than £5, e.g. £12.

Let's Investigate

- Ask children to comment on the strengths and limitations of each of their methods. Which one do they prefer to use? Why?
- When the children have used a range of different methods to work out the answer to the calculation, arrange them into pairs. Children discuss and compare the different methods used.

112

Page 3

The Puzzler

- Ask the children to solve this puzzle. How many different answers are there?

Looking for Patterns

- Once the children have completed the activity, arrange them into pairs or groups to compare and discuss their methods of working. Which method(s) were the most effective?
- Tell the children the value of one or more of the missing digits (see Answers).

Page 4

Let's Investigate

- Children investigate whether the same applies for other magic squares, for example:

5	22	18
28	15	2
12	8	25

9	6	3	16
4	15	10	5
14	1	8	11
7	12	13	2

17	24	1	8	15
23	5	7	14	16
4	6	13	20	22
10	12	19	21	3
11	18	25	2	9

Looking for Patterns

- You may wish to provide the children with 2 cm squared paper on which to record their calculations.
- Provide the children with a set of dominoes so that they can rotate and position them to form the calculations.

AfL

- How did you find the solution to this puzzle? How did you start?
- What different combinations of things were you able to buy? Which combination bought you the most items? Which combination gave you the least number of items?
- Explain the different methods you used to work out the answer to the calculation. Which of these methods do you prefer? Why?
- What patterns did you look out for? How did these patterns help you?
- What can you tell me about magic squares?

Issue 6 – Addition

Answers

Page 1

The Puzzler
A = 9, H = 1, P = 2 and R = 8
(991 + 991 = 1982)
Explanations will vary.

Money Matters

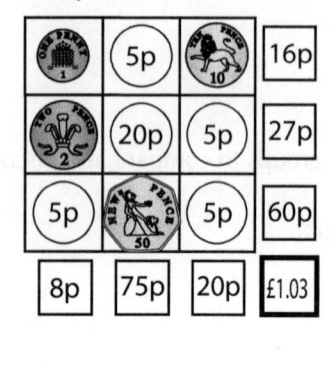

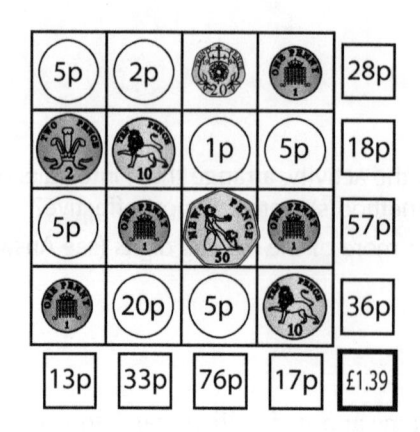

			16p
1p	5p	10p	16p
2p	20p	5p	27p
5p	50p	5p	60p
8p	75p	20p	£1.03

			28p
5p	2p	20p	1p	28p
2p	10p	1p	5p	18p
5p	1p	50p	1p	57p
1p	20p	5p	10p	36p
13p	33p	76p	17p	£1.39

Page 2

Money Matters
Combinations of items will vary.

Let's Investigate
Calculating methods will vary.

Page 3

The Puzzler

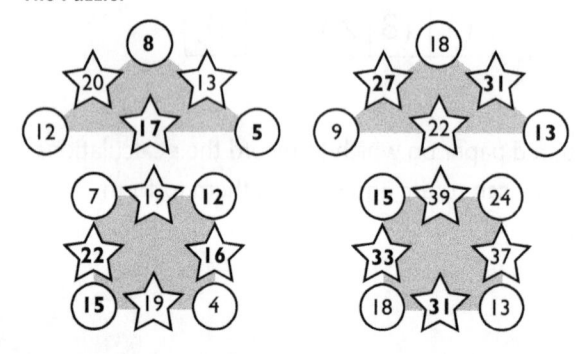

Star 1: 8, 20, 13, 12, 17, 5, 7, 19, 12, 22, 16, 15, 19, 4
Star 2: 18, 27, 31, 9, 22, 13, 15, 39, 24, 33, 37, 18, 31, 13

Looking for Patterns
32 + 16 = 48   25 + 49 = 74
67 + 25 = 92   73 + 34 = 107
Explanations will vary.

Page 4

Let's Investigate
The magic number for the 3 × 3 square is 15.
For each change ① to ⑥ the square is still magic.
The magic number for the 4 × 4 square is 34.
For each change ① to ⑥ the square is still magic.

Looking for Patterns

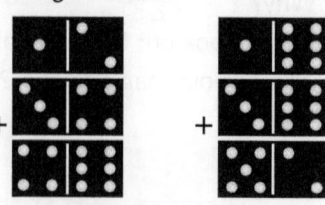

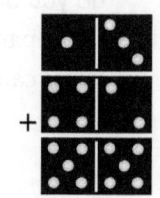

Other solutions are possible.

Inquisitive ant

calculation
The process, or a step in the process, of working out the answer to a mathematical problem involving numbers.

Issue 7

Subtraction

Prerequisites for learning

- Identify patterns and relationships involving numbers
- Recall and use addition and subtraction facts to 20 fluently, and derive and use related facts up to 100
- Add and subtract numbers mentally, including:
  - a two- or three-digit number and ones
  - a two- or three-digit number and tens
  - two two-digit numbers
  - adding three one-digit numbers
  - a three-digit number and hundreds
- Add and subtract numbers with up to three digits, using formal written methods of columnar addition and subtraction

Resources

pencil and paper
Resource sheet 2: My notes (optional)
Resource sheet 3: Pupil self assessment booklet (optional)
calculator (optional)

Teaching support

Page 1

Let's Investigate

- Children investigate what happens if you start with a two-digit or a four-digit number.

Let's Investigate

- This method of subtraction involves partitioning the minuend (the number from which another number is to be subtracted) into a multiple of 100 and the remainder, e.g. 386 = 300 + 86, and then subtracting the subtrahend (the number that is to be subtracted from the minuend) from the multiple of 100, i.e. 300 − 178 = 122. The difference is then added to the remainder, i.e. 86 + 122 = 208.

- Discuss with the children how the minuend can be partitioned in other ways, for example:

```
   386         200 (+ 186)        186
 − 178        − 178             +  22
 _____        _____            _____
                  22               208
                                    1
```

- Children use the method for subtraction calculations involving combinations of three-digit and four-digit numbers, e.g. 2436 − 857 and 3452 − 1768.

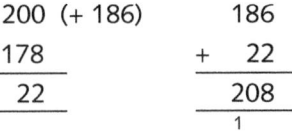

Page 2

Let's Investigate

- Ensure the children realise that this activity requires them to include all the subtraction number facts for 0, 1, 2, 3, 4, 5, 6, 7, 8, 9 and 10.
- Discuss with the children the patterns they notice in the subtraction number facts for 5. This will help them to identify the subtraction facts for other numbers.
- Although all the subtraction number facts for 5 are included in the issue as an example, if necessary work with the children to identify all the subtraction facts for another number, e.g. 8 (see Answers).
- Tell the children that altogether there are 66 subtraction number facts to 10.

Let's Investigate

- Ask children to comment on the strengths and limitations of each of their methods. Which one do they prefer to use? Why?
- When the children have used a range of different methods to work out the answer to the calculation, arrange them into pairs. Children discuss and compare the different methods used.

115

Issue 7 – Subtraction

Page 3

The Puzzler
- Children write a pentagonal or octagonal subtraction puzzle for a friend to solve.

In the Past
- Ensure children understand the shape of an amphitheatre before starting this activity.

Page 4

Let's Investigate
- Tell the children the following:
  - There are four different calculations possible with a difference of 9.
  - There are six different calculations possible with a difference of 10.
  - There are four different calculations possible with a difference of 11.
  - There are two different calculations possible with a difference of 12.

Let's Investigate
- Tell the children there are 20 other different calculations possible.

The Puzzler
- Ensure children realise that there are 10 digits (0–9)
- Ask the children to write as many different calculations as they can.
- What if the calculation was M – I – N – U – S = 0? How many different calculations could they make? (e.g. 6 – 0 – 1 – 2 – 3 = 0)

AfL

- What are all the subtraction number facts for 6? What do you notice about these facts? What relationships can you see?
- What is similar about these two facts?
- What patterns did you notice in your answers?
- Explain this method of subtraction to me. Is it a method that you might use? Why? Why not?
- Explain the different methods you used to work out the answer to the calculation. Which of these methods do you prefer? Why?
- How did you find the solution to this puzzle? How did you start?
- How many different calculations were you able to find? Is this all of them? How do you know?

Answers

Page 1

Let's Investigate
The answer is always 18.

Let's Investigate

```
  573        500         73
- 495      - 495        +  5
              5         78

  743        700         43
- 256      - 256       + 444
            444        487

  359        300         59
- 168      - 168       + 132
            132        191

  835        800         35
- 347      - 347       + 453
            453        488

  773        700         73
- 684      - 684       +  16
             16         89
```

Calculations and explanations will vary.

Page 2

Let's Investigate
There are 66 subtraction number facts up to, and including, 10.

0 – 0 = 0	1 – 0 = 1	2 – 0 = 2	3 – 0 = 3
	1 – 1 = 0	2 – 1 = 1	3 – 1 = 2
		2 – 2 = 0	3 – 2 = 1
			3 – 3 = 0

4 – 0 = 4	5 – 0 = 5	6 – 0 = 6	7 – 0 = 7
4 – 1 = 3	5 – 1 = 4	6 – 1 = 5	7 – 1 = 6
4 – 2 = 2	5 – 2 = 3	6 – 2 = 4	7 – 2 = 5
4 – 3 = 1	5 – 3 = 2	6 – 3 = 3	7 – 3 = 4
4 – 4 = 0	5 – 4 = 1	6 – 4 = 2	7 – 4 = 3
	5 – 5 = 0	6 – 5 = 1	7 – 5 = 2
		6 – 6 = 0	7 – 6 = 1
			7 – 7 = 0

8 – 0 = 8	9 – 0 = 9	10 – 0 = 10
8 – 1 = 7	9 – 1 = 8	10 – 1 = 9
8 – 2 = 6	9 – 2 = 7	10 – 2 = 8
8 – 3 = 5	9 – 3 = 6	10 – 3 = 7
8 – 4 = 4	9 – 4 = 5	10 – 4 = 6
8 – 5 = 3	9 – 5 = 4	10 – 5 = 5
8 – 6 = 2	9 – 6 = 3	10 – 6 = 4
8 – 7 = 1	9 – 7 = 2	10 – 7 = 3
8 – 8 = 0	9 – 8 = 1	10 – 8 = 2
	9 – 9 = 0	10 – 9 = 1
		10 – 10 = 0

Let's Investigate
Calculating method will vary.

Page 3

The Puzzler

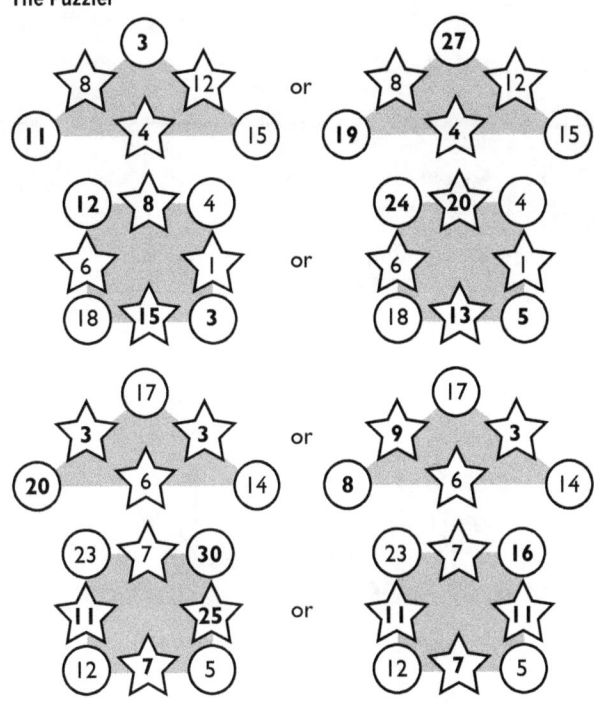

In the Past
The Coliseum's length is 33 m greater than its width.
The central arena's length is 33 m greater than its width.
The width of the seating that runs around the arena is 51 m.

Page 4

Let's Investigate
Calculations with a difference of 9:
98 – 89   97 – 88
88 – 79   87 – 78
Calculations with a difference of 10:
99 – 89   98 – 88
97 – 87   89 – 79
88 – 78   87 – 77
Calculations with a difference of 11:
99 – 88   98 – 87
89 – 78   88 – 77
Calculations with a difference of 12:
99 – 87   89 – 77

Let's Investigate

99 – 98 = 1	98 – 97 = 1	97 – 89 = 8
99 – 97 = 2	98 – 79 = 19	97 – 79 = 18
99 – 79 = 20	98 – 78 = 20	97 – 78 = 19
99 – 78 = 21	98 – 77 = 21	97 – 77 = 20
99 – 77 = 22		

89 – 88 = 1	88 – 87 = 1	87 – 79 = 8
89 – 87 = 2		

79 – 78 = 1   78 – 77 = 1
79 – 77 = 2

The Puzzler
9 – 0 – 1 – 2 – 3 = 3
Other solutions are possible.

Inquisitive ant

inverse relationship
The inverse of something is the opposite. The inverse of subtraction is addition (and vice versa). The inverse of multiplication is division (and vice versa).

Issue 8

Subtraction

Prerequisites for learning

- Identify patterns and relationships involving numbers
- Recall and use addition and subtraction facts to 20 fluently, and derive and use related facts up to 100
- Add and subtract numbers mentally, including:
  - a two- or three-digit number and ones
  - a two- or three-digit number and tens
  - two two-digit numbers
  - adding three one-digit numbers
  - a three-digit number and hundreds
- Add and subtract numbers with up to three digits, using formal written methods of columnar addition and subtraction

Resources

pencil and paper
Resource sheet 2: My notes (optional)
Resource sheet 3: Pupil self assessment booklet (optional)
Resource sheet 9: 2 cm squared paper (optional)
ruler
set of dominoes (optional)
set of 1–9 digit cards
1–6 dice (optional)
scissors (optional)

Teaching support

Page 1

The Puzzler

- ATT is a number between 100 and 108, as subtracting a one-digit number from it gives an answer that is a two-digit number. It is a three-digit number with the same units and tens digit. So A = 1 and T = 0. Subtracting a one-digit number from 100 must give an answer between 91 and 99, so S = 9.

Let's Investigate

- There are 24 different calculations possible, with 20 different answers (see Answers).
- Can the children find all 24 different calculations?
- Once children have completed the activity, arrange them into pairs to compare and discuss their calculations.

Page 2

Let's Investigate

- Once children have completed the activity, arrange them into pairs to compare and discuss their calculations.

What's the Problem?

- In 35 minutes the bag leaked 35 × 12 = 420 ml.
  1 litre – 420 ml = 580 ml.

Money Matters

- During the day Orson lost or spent £5.47 altogether. This was half the amount he had started with.

Page 3

Looking for Patterns

- You may wish to provide the children with 2 cm squared paper on which to record their calculations.
- Provide the children with a set of dominoes so that they can rotate and position them to form the calculations.

Let's Investigate

- Allow the children to work in pairs.

What's the Problem?

- If they both start off showing 12 o'clock, after three days the clock that loses one hour every 12 hours will show 6 o'clock, and the one that loses two hours every 12 hours will show 12 o'clock. After six days they will both show 12 o'clock.
- Tell children to draw a table showing the time that each clock shows after $\frac{1}{2}$ day, 1 day, $1\frac{1}{2}$ days, 2 days, etc. if they both start off at 12 o'clock.

Page 4

The Puzzler

- Provide the children with a 1–6 dice.
- Suggest the children draw each of the three diagrams onto 2 cm squared paper, cut them out, fold them into cubes and draw on the missing dots.

Around the World

- Encourage the children to write comparative statements involving multiplication and division (scaling), for example: "Almost 10 times as many people live in London than live in Stockholm" "100 times as many people live in Riga than live in Valletta".
- Once children have completed the activity, arrange them into pairs to compare and discuss their statements.

AfL

- How many different calculations were you able to write? Do you think there are any more? Why?
- How did you find the solution to this puzzle / answer to this problem?
- Tell me one of your statements. What calculations did you have to do?

Issue 8 – Subtraction

Answers

Page 1

The Puzzler
A = 1, T = 0 and S = 9
(100 – 9 = 91)
Explanations will vary.

Let's Investigate
The 20 different numbers that can be made are: 8, 9, 11, 12, 17, 18, 19, 21, 22, 23, 27, 28, 29, 31, 32, 33, 38, 39, 41, 42.

The 24 calculations are:
12 – 4 = 8	12 – 3 = 9	13 – 4 = 9
13 – 2 = 11	14 – 3 = 11	14 – 2 = 12
21 – 4 = 17	21 – 3 = 18	23 – 4 = 19
24 – 3 = 21	23 – 1 = 22	24 – 1 = 23
31 – 4 = 27	32 – 4 = 28	31 – 2 = 29
32 – 1 = 31	34 – 2 = 32	34 – 1 = 33
41 – 3 = 38	41 – 2 = 39	42 – 3 = 39
42 – 1 = 41	43 – 2 = 41	43 – 1 = 42

Page 2

Let's Investigate
Pairs of numbers with a difference of 6 are:

1		18		22		2		21	15
7		12		16		8		3	9

Pairs of numbers with a difference of 3 are:

14	11		7
			10

Pairs of numbers with a difference of 4 are:

23		10		20	24
19		6			

Pairs of numbers with a difference of 5 are:

7	12		8
			13

What's the Problem?
Yes, Lucy did get the goldfish home safely. There was still 580 ml of water left in the bag when she arrived home.

Money Matters
Orson had £10.94 when he set off to school.

Page 3

Looking for Patterns

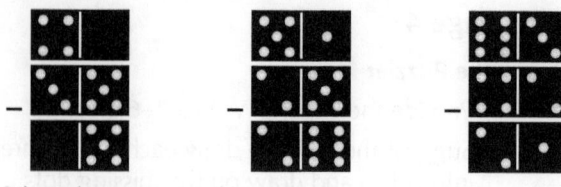

Other solutions are possible.

Let's Investigate
Calculations will vary.

What's the Problem?
It will be six days before both clocks show the same time again.

Page 4

The Puzzler

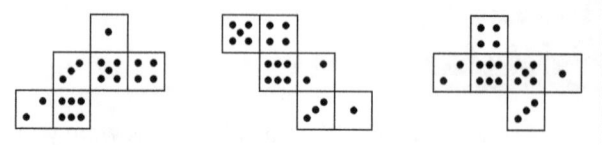

Around the World
Statements will vary.

Inquisitive ant

prediction
A statement or opinion about what the result of something will be.

Issue 9
Multiplication

Prerequisites for learning

- Recall and use multiplication and division facts for the 2, 3, 4, 5, 8 and 10 multiplication tables
- Recognise multiples of 2, 3, 4, 5, 8 and 10
- Write and calculate mathematical statements for multiplication, including for two-digit numbers multiplied by one-digit numbers, using mental and progressing to formal written methods
- Use diagrams to sort data and objects using more than one criterion
- Solve logic puzzles

Resources

pencil and paper
Resource sheet 2: My notes (optional)
Resource sheet 3: Pupil self assessment booklet (optional)
calculator (optional)

Teaching support

Page 1

The Puzzler
- Ensure children understand that the term "product" refers to the result of a multiplication calculation.
- Tell the children the location of one or two of the digits on the grid (see Answers).

Money Matters
- Suggest the children draw a table to help them find the solutions to the problems, for example:

Hour	Amount	Total
9:00–10:00	£10	£10
10:00–11:00	£20	£30
11:00–12:00	£40	£70
12:00–1:00	£80	£150
1:00–2:00	£160	£310
2:00–3:00	£320	£630
3:00–4:00	£640	£1270
4:00–5:00	£1280	£2550
5:00–6:00	£2560	£5110

The Puzzler
- Suggest the children rewrite the calculation as $CAT = \frac{AAA}{3}$. So AAA divided by 3 gives a three-digit whole number answer with three different digits, the middle one of which is the same as A. Substitute different digit values for A to find the answer.
- Tell the children the value of one of the letters (see Answers).
- Allow the children to use a calculator.

Page 2

The Puzzler
- Ensure children are able to interpret the diagram.
- Also ensure that the children write down their predictions before completing the diagram.
- You may wish to ask the children to do this activity in pairs so that they can share their predictions and discuss their reasoning.

What's the Problem?

- Suggest the children record the information in a table, for example:

No. of each animal	1	2	3	4	5	6	7	8	9	10	11	12	13	14	15
No. of sheeps' ears and tails	3	6	9	12	15	18	21	24	27	30	33	36	39	42	45
No. of chickens' legs	2	4	6	8	10	12	14	16	18	20	22	24	26	28	30

Page 3

Let's Investigate

- Physically work through an example with the children.
- The important aspect of this investigation is the opinion children have of this method and the method's strengths and limitations.

Page 4

The Language of Maths

- Although this problem involves no actual mathematics, it does require the children to problem solve, reason and think logically.

What's the Problem?

- Ahmed has to pay six gold coins each day to use the chest, but the number of coins he has stays constant each day. Therefore, the increase in the number of coins in the chest each night must be six. That means that six coins must be put into the chest each evening. As Ahmed pays Sinbad before he puts his coins in the chest, he must start with 12 gold coins each evening.
- Suggest the children use trial and improvement.
- How much would Ahmed have had after six nights if he had only had to pay Sinbad five gold coins each night? (138 gold coins)

Let's Investigate

- Ask children to comment on the strengths and limitations of each of their methods. Which one do they prefer to use? Why?
- When the children have used a range of different methods to work out the answer to the calculation, arrange them into pairs. Children discuss and compare the different methods used.

AfL

- How did you work out the solution to this puzzle? Where did you begin?
- What strategies did you use to help you?
- Explain your diagram to me. Did this match with your earlier predictions?
- What do you think of the finger method of multiplication?
- Explain the different methods you used to work out the answer to the calculation. Which of these methods do you prefer? Why?

Answers

Page 1

The Puzzler

4	2	6
1	8	3
9	5	7

Other arrangements are possible.

Money Matters
The Big Read took £1280 between 4 o'clock and 5 o'clock.
The Big Read took £5110 during the entire day.

The Puzzler
C = 1; A = 4 and T = 8
(3 × 148 = 444)
Explanations will vary.

Page 2

The Puzzler

	Multiple of 8	Multiple of 10
Multiple of 3	24	30
Multiple of 4	8, 16, 24, 32, 40, 48	40
Multiple of 5	40	10, 20, 30, 40, 50, 60, 70, 80, 90, 100, 110, 120

What's the Problem?
There are 10 sheep and 15 chickens.

Page 3

Let's Investigate
Explanations will vary.

Page 4

The Language of Maths
After 13 days the casket was half full of gold coins.

What's the Problem?
Ahmed started with 12 gold coins.

Let's Investigate
172 × 4 = 688
Calculating methods will vary.

Inquisitive ant

scaling
The method of multiplication in which a given amount is increased by a scale factor, e.g. doubling is a scale factor of 2; trebling is a scale factor of 3.

Issue 10
Multiplication

Prerequisites for learning

- Identify patterns and relationships involving numbers
- Recall and use multiplication and division facts for the 2, 3, 4, 5, 8 and 10 multiplication tables
- Recognise multiples of 2, 3, 4, 5, 8 and 10
- Write and calculate mathematical statements for multiplication, including for two-digit numbers multiplied by one-digit numbers, using mental and progressing to formal written methods
- Solve logic puzzles

Resources

pencil and paper
Resource sheet 2: My notes (optional)
Resource sheet 3: Pupil self assessment booklet (optional)
calculator

Teaching support

Page 1

The Puzzler

- As the product is a three-digit number, LEG must be a number less than 333, otherwise the product would be larger than 999. So L must be 1, 2 or 3.

  If L = 2, the product of 3 × G must end in 2, so G would be 4 and E would be 6, 7 or 8 (because 3 × 2E4 = EE2). None of which fit.

  If L = 3, 3 × G must end in 3, so G would be 1 and E would be 9 (because 3 × 3E1 = EE3). This doesn't fit, so L must equal 1, and G must equal 7 (because 3 × G = ☐1). Trial and improvement gives E = 4.

- Tell the children the value of one of the letters (see Answers).
- Allow the children to use a calculator.

Looking for Patterns

- Children investigate patterns in other sets of calculations, for example:

  37 037 × 3 = 111 111   3367 × 33 = 111 111
  37 037 × 6 = 222 222   3367 × 66 = 222 222
  37 037 × 9 = 333 333   3367 × 99 = 333 333

Page 2

In the Past

- Ensure children understand the duplation table before working independently on this activity.
- Children draw a duplation table for the number 12 and use it to answer related multiplication calculations, e.g. 12 × 7, 12 × 16, 12 × 19. They check their answers using their own method or a calculator.

Looking for Patterns

- The pattern is as follows:

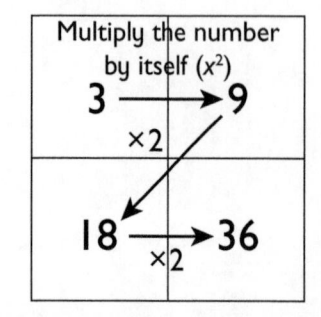

Multiply the number by itself (x^2)		Multiply the number by itself (x^2)
3 →9 ×2 ↘ 18 →36 ×2	or	3 →9 ↓×4 18← 36 ÷2

124

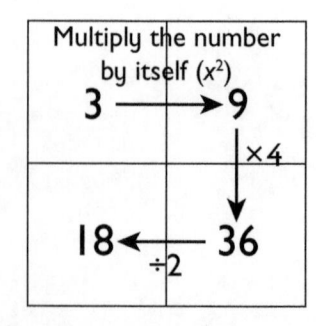

Page 3

The Puzzler

- Ask the children whether or not they can complete each of the puzzles if the numbers next to each other were added together to equal the number in the box above. Is there more than one solution to each of the puzzles?

Let's Investigate

- Ask the children to find the largest product they can make for a two-digit number multiplied by a two-digit number calculation, i.e. 41 × 32 = 1312

The Puzzler

- Ensure children are familiar with the terms "consecutive whole numbers" and "product" (Inquisitive ant).
- One method of finding a solution to this puzzle is to realise that 336 can be broken down into the prime factors 2 × 2 × 2 × 2 × 3 × 7. The numbers it is possible to make with these are 2, 3, 4, 6, 7, 8, 12, 14, 16, etc. Numbers 6, 7 and 8 fit the question.
- Children use trial and improvement to find the correct numbers.

Page 4

Let's Investigate

- Tell the children that there are six different calculations possible for each number sentence.

Technology Today

- The 24-month contract is slightly cheaper, £402 as opposed to £408, but ties Sam in for twice the length of time. Something cheaper may be available in 12 months time, and probably will as technology keeps getting cheaper.

AfL

- How did you solve that puzzle? What did you do to start? What things did you think of?
- Tell me what you think of the duplation method of multiplication. What is good about this method? What is not so good? Is it a method that you might use? Why? Why not? When?
- What patterns do you notice in this puzzle?

Answers

Page 1

The Puzzler
E = 4, G = 7 and L = 1
(3 × 147 = 441)
Explanations will vary.

Looking for Patterns
37 × 3 = 111 and 1 + 1 + 1 = 3
37 × 6 = 222 and 2 + 2 + 2 = 6
37 × 9 = 333 and 3 + 3 + 3 = 9
37 × 12 = 444 and 4 + 4 + 4 = 12
37 × 15 = 555 and 5 + 5 + 5 = 15
37 × 18 = 666 and 6 + 6 + 6 = 18

When you add together the digits in each answer:
- the total is the same number as the number you originally multiplied 37 by, for example:

37 × **3** = 111 and 1 + 1 + 1 = **3**

- the numbers are all multiples of 3.

37 × 3 = 111 and 1 + 1 + 1 = 3
37 × 33 = 1221 and 1 + 2 + 2 + 1 = 6
37 × 333 = 12 321 and 1 + 2 + 3 + 2 + 1 = 9
37 × 3333 = 123 321 and 1 + 2 + 3 + 3 + 2 + 1 = 12
37 × 33 333 = 1 233 321 and 1 + 2 + 3 + 3 + 3 + 2 + 1 = 15
37 × 333 333 = 12 333 321 and 1 + 2 + 3 + 3 + 3 + 3 + 2 + 1 = 18

When you add together the digits in each answer:
- the total is the same as the total of the digits in the number you originally multiplied 37 by, for example:
37 × 333 333 = 12 333 321 and 1 + 2 + 3 + 3 + 3 + 3 + 2 + 1 = **18**

3 + 3 + 3 + 3 + 3 + 3 = **18**

- the numbers are all multiples of 3.

Page 2

In the Past
17 × 3 = 17 + 34 = 51
17 × 6 = 34 + 68 = 102
17 × 9 = 17 + 136 = 153
17 × 12 = 68 + 136 = 204

Looking for Patterns

6	36
72	144

8	64
128	256

9	81
162	324

Page 3

The Puzzler

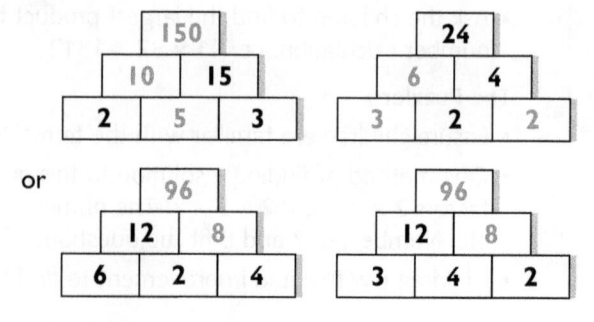

```
      150
   10     15
  2    5    3
```

```
      24
    6     4
  3    2    2
```

or

```
      96
   12     8
  6    2    4
```

```
      96
   12     8
  3    4    2
```

Let's Investigate
4 × 32 = 128
4 × 321 = 1284

The Puzzler
6 × 7 × 8 = 366
Explanations will vary.

Page 4

Let's Investigate

2 × 60 = 120   2 × 72 = 144
3 × 40 = 120   3 × 48 = 144
4 × 30 = 120   4 × 36 = 144
5 × 24 = 120   6 × 24 = 144
6 × 20 = 120   8 × 18 = 144
8 × 15 = 120   9 × 16 = 144

Technology Today
Advice and reasoning will vary.

Inquisitive ant

product
The result of the multiplication of two or more numbers.

Issue 11
Division

Prerequisites for learning

- Identify patterns and relationships involving numbers
- Recall and use multiplication and division facts for the 2, 3, 4, 5, 8 and 10 multiplication tables
- Recognise multiples of 2, 3, 4, 5, 8 and 10, and begin to recognise multiples of 6 and 7
- Write and calculate mathematical statements for division, including for two-digit numbers by one-digit numbers, using mental and progressing to formal written methods
- Begin to understand that in a division situation a remainder (abbreviated as 'r') is the amount left over after an equal sharing or grouping has been completed

Resources

pencil and paper
Resource sheet 2: My notes (optional)
Resource sheet 3: Pupil self assessment booklet (optional)
30 counters (optional)
calculator (optional)

Teaching support

Page 1

Looking for Patterns

- If necessary work through the first set of numbers with the children, for example:
  9 and 81
  $81 \div 9 = 9$
  $18 - 9 = 9$
- Can the children find other pairs of one-digit and two-digit numbers that have the same relationship?

Looking for Patterns

- Provide the children with 30 counters to assist them with working out the solution to the problem.

Page 2

The Arts Roundup

- This activity involves the children calculating with numbers that they may not yet be familiar with. However the aim of this activity is to encourage children to use known multiplication and division facts for the 8 multiplication table in order to solve the problem.
- To cover the total cost of production, each advert costs £120 000 (£960 000 ÷ 8).
- You may wish to allow the children to use a calculator for this activity.

The Puzzler

- Ensure children understand the description: "The sum of the two digits is 16." For example, the sum of the digits of the number 54 is 9, i.e. 5 + 4 = 9.
- Children write a similar puzzle for a friend to solve.

Looking for Patterns

- You may wish to introduce the children to the concept of common multiples as a way of helping them work out the answer to this problem. (84 is the only common multiple of 3, 4 and 7 less than 100.)

Page 3

Let's Investigate
- Ask the children to investigate what happens if you start with a four-digit number. (No similar pattern occurs.)

What's the Problem?
- If all the heads belonged to men, there would only be 120 legs. However, there are 40 more legs than this. Each horse has two more legs than a man, so there must be 40 ÷ 2 horses = 20 horses for the number of heads and legs to be correct.

Page 4

What's the Problem?
- One method of solving this problem is to find the lowest common multiple of 2, 3, 4, 5 and 6 and add one.
- What is the smallest three-digit number that has a remainder of 1 when divided by 3, 5 and 8? (121)
- An important aspect of this activity is the explanation children give for discovering the number. Therefore, when children have finished, arrange them into pairs or groups to compare and discuss their methods.

Let's Investigate
- Ask children to comment on the strengths and limitations of each of their methods. Which one do they prefer to use? Why?
- When the children have used a range of different methods to work out the answer to the calculation, arrange them into pairs. Children discuss and compare the different methods used.

Sports Update
- The first flag is at the Start line and the eleventh at the Finish, so the sixth flag is half-way along the course.
- Suggest children draw a diagram to help them, for example:

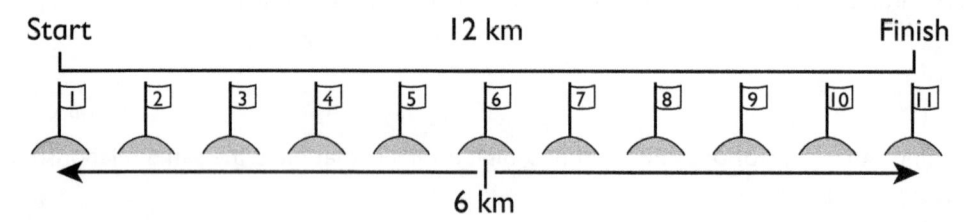

AfL

- What patterns did you notice? How did this help you?
- How did you work out the answer to this problem? Did you use a diagram to help you? How did that help?
- Explain the different methods you used to work out the answer to the calculation. Which of these methods do you prefer? Why?

Answers

Page 1

Looking for Patterns
9 and 81   3 and 72   2 and 94
81 ÷ 9 = 9   72 ÷ 3 = 24   94 ÷ 2 = 47
18 − 9 = 9   27 − 3 = 24   49 − 2 = 47
The answers to both calculations are the same.

Looking for Patterns
Emily picked up 21 counters.

Page 2

The Arts Roundup
Each advert costs £120 000.

The Puzzler
The number in the box is 97.

Looking for Patterns
There are 84 apples in the crate.

Page 3

Let's Investigate
The final answer is always 11.

What's the Problem?
There are 20 horses.

Page 4

What's the Problem?
61
Explanations will vary.

Let's Investigate
Calculating methods will vary.

Sports Update
When Omar passes the sixth flag he has run 6 km.

Inquisitive ant

factor
A whole number that will divide exactly into another whole number.

Issue 12

Division

Prerequisites for learning

- Identify patterns and relationships involving numbers
- Recall and use multiplication and division facts for the 2, 3, 4, 5, 8 and 10 multiplication tables
- Recognise multiples of 2, 3, 4, 5, 8 and 10, and begin to recognise multiples of 6 and 7
- Identify the doubles of two-digit numbers and use these to calculate doubles of multiples of 10
- Multiply and divide numbers to 1000 by 10 (whole-number answers)
- Write and calculate mathematical statements for division, including for two-digit numbers by one-digit numbers, using mental and progressing to formal written methods
- Begin to understand that in a division situation a remainder (abbreviated as 'r') is the amount left over after an equal sharing or grouping has been completed

Resources

pencil and paper
Resource sheet 2: My notes (optional)
Resource sheet 3: Pupil self assessment booklet (optional)
counters and matchsticks, or similar (optional)
set of 0–9 digit cards

Teaching support

Page 1

What's the Problem?

- If all the heads belonged to people, there would be only 26 × 2 = 52 legs. However, there are 76 − 52 = 24 extra legs present. Each dog has two more legs than a person, so there are $\frac{24}{2}$ = 12 dogs present.
- Use counters and matchsticks (or similar) to represent the heads and legs, and discuss the above explanation graphically.

The Arts Roundup

- Suggest children draw a table to find out the number of members of BADS, for example:

	Size of groups				
Number of members	2	3	4	5	6
50	✓	✗	✗	✓	✗
51	✗	✓	✗	✗	✗
52	✓	✗	✓	✗	✗
53	✗	✗	✗	✗	✗
54	✓	✓	✗	✗	✓
55	✗	✗	✗	✓	✗
56	✓	✗	✓	✗	✗

- What patterns do they identify? Can they use these patterns to work out the answer to the problem?

Focus on Science

- Ensure children know that one tonne is equivalent to 1000 kg.

Page 2

What's the Problem?
- Although this problem involves no actual mathematical calculation, it does require the children to problem solve, reason and think logically.
- The second sock that she chooses may be a different colour to the first, but the third is certain to match either the first or the second.

Let's Investigate
- Remind the children that a square is a type of rectangle and therefore should be included in this investigation.
- Do not give the children counters. Can they still make all the multiplication calculations possible with products from 10 to 24?
- What do the children notice about the numbers that make squares?

Looking for Patterns
- Suggest the children try each of the multiples of 8 in ascending size.

Page 3

Let's Investigate
- Ensure children realise that the numbers 2, 4, 6, 8, 10, 12, 14, 16, … are referred to as *multiples of 2*; and that the numbers 4, 8, 12, 16, 20, 24, 28, 32, … are referred to as *multiples of 4*.

Looking for Patterns
- The number must be even, so it must be an even multiple of 7, i.e. a multiple of 14. The smallest number that is divisible by both 14 and 6 is 42, so the number is a multiple of 42. 42 is not divisible by 8, nor is 84, nor 126; however, 168 is.

Looking for Patterns
- The example in the issue uses nine out of the ten cards to show different multiples of 6. Encourage the children to use as many of the digit cards as possible when making multiples of any number. Can they use all ten cards?

Page 4

At Home
- Once the children have completed the investigation, ensure that there is an opportunity in class for pairs or groups of children to discuss their results.

Let's Investigate
- Ask the children to write about the limitations of this method.

AfL

- Explain to me how you worked out the answer to this problem.
- Tell me some of your calculations. What do you notice about them? How else could you describe this array? What does this mean?
- What patterns did you notice in the units digits?
- Is there another way that you could arrange these cards to show me some other multiples of …? Is there a way you could use even more of the cards?
- What did you find out about the way different food and other items are packaged? Why do you think this is?
- What do you think of this method for dividing a number by 5? When it is not useful?

Issue 12 — Division

Answers

Page 1

What's the Problem?
There are 14 people and 12 dogs in the dog training class.

The Arts Roundup
There are 60 members of BADS.

Focus on Science
Peter's supply of logs will last for 75 days.

Page 2

What's the Problem?
The minimum number of socks that Kylie needs to take out of the drawer in order to be certain of getting a matching pair is three socks.

Let's Investigate
10: 1 × 10, 2 × 5
11: 1 × 11
13: 1 × 13
14: 1 × 14, 2 × 7
15: 1 × 15, 3 × 5
16: 1 × 16, 2 × 8, 4 × 4
17: 1 × 17
18: 1 × 18, 2 × 9, 3 × 6
19: 1 × 19
20: 1 × 20, 2 × 10, 4 × 5
21: 1 × 21, 3 × 7
22: 1 × 22, 2 × 11
23: 1 × 23
24: 1 × 24, 2 × 12, 3 × 8, 4 × 6

Looking for Patterns
34

Page 3

Let's Investigate
The patterns of the 2 units digit for the multiples of 2 and 4 are:
②, 4, 6, 8, 10, 1②, 14, 16, 18, 20, 2②, 24, 26, 28, …
4, 8, 1②, 16, 20, 24, 28, 3②, 36, 40, 44, 48, 5②, 56, …

2 as the units digit appears in every fifth number in both sequences. Similar patterns occur for the other units digits.

Patterns for the multiples of 3 and 6, and 4 and 8 will vary.

Looking for Patterns
168

Looking for Patterns
Results of the investigation will vary.

Page 4

At Home
Results of the investigation will vary.

Let's Investigate
95 ÷ 5 = 19; 180 ÷ 5 = 36; 425 ÷ 5 = 85; 715 ÷ 5 = 143.
The method works for all multiples of 5.

Inquisitive ant

multiple
A number that can be divided exactly by another smaller number without a remainder.

Issue 13

Mixed operations

Prerequisites for learning

- Identify patterns and relationships involving numbers
- Recall and use addition and subtraction facts to 20 fluently, and derive and use related facts up to 100
- Add and subtract numbers mentally, including:
  - a two- or three-digit number and ones
  - a two- or three-digit number and tens
  - two two-digit numbers
  - adding three one-digit numbers
  - a three-digit number and hundreds
- Add and subtract numbers with up to three digits, using formal written methods of columnar addition and subtraction
- Recall and use multiplication and division facts for the 2, 3, 4, 5, 8 and 10 multiplication tables
- Write and calculate mathematical statements for multiplication and division, including for two-digit numbers by one-digit numbers, using mental and progressing to formal written methods
- Begin to understand and use the order of operations, including brackets

Resources

pencil and paper
Resource sheet 2: My notes (optional)
Resource sheet 3: Pupil self assessment booklet (optional)
computer with Internet access

Teaching support

Page 1

Money Matters

- Suggest the children create their own monetary system. What will they use as currency? What denominations will they have? Ensure the children give explanations for their reasoning.

What's the Problem?

- Ensure children understand and are able to use the order of operations, including brackets.

Page 2

Money Matters

- Encourage the children to use fractions to describe how house prices have changed over the past 12 months / compared with those in a different area nearby.
- Encourage the children to use percentages to describe how house prices have changed / compare.

At Home

- Prior to children working independently on this activity, you may need to discuss with them what constitutes a "big saving".
- Once the children have completed the activity, ensure that there is an opportunity in class for pairs or groups of children to discuss their results.

The Language of Maths

- Ensure children are familiar with square numbers to at least 10 × 10.
- Once the children have completed the activity, provide an opportunity for them to talk about, and compare, their methods for calculating the squares of the numbers.

Issue 13 – Mixed operations

Page 3

Money Matters

- It is recommended that children work in pairs for this activity so that they can share ideas and discuss their reasoning.
- The most important aspect of this activity is the reasoning that children offer when justifying how much pocket money they think they should receive.

Sports Update

- Suggest the children solve the problem by working backwards, for example:

  Sunday: 40 min
  Saturday: 40 min – 5 min = 35 min
  Friday: 35 min – 5 min = 30 min
  Thursday: 30 min – 5 min = 25 min
  Wednesday: 25 min – 5 min = 20 min
  Tuesday: 20 min – 5 min = 15 min
  Monday: 15 min – 5 min = 10 min

The Puzzler

- If children experience difficulty starting the puzzle, draw their attention to the end of the puzzle. Explain that as the final answer is 23 and the operation is + 9, the number that belongs in the square before + 9 must equal 23, i.e. ☐ + 9 = 23. Therefore the missing number must be 14.
- You may also wish to remind children of the inverse relationship between addition and subtraction.

Page 4

The Puzzler

- If children experience difficulty starting the puzzle, draw their attention to the start of the puzzle. Ask questions such as: "Is 17 more or less than 12? How many more? So, what do you need to add to 12 to make 17?"
- In both puzzles, there are two empty boxes where two different operations are possible (see Answers). Can the children find them both?

What's the Problem?

- Suggest the children draw a table to work out the solution to the problem, for example:

Bus stop number	Number of people who get on the bus	Total number of people on the bus
1	1	1
2	2	3
3	3	6
4	4	10
5	5	15
6	6	21

AfL

- How do you know that this is the correct answer to this calculation?
- What can you tell me about house prices in your neighbourhood?
- Is there a big difference between the normal price of a supermarket item and its sale price? Which types of items seem to offer a "big saving"?
- Tell me the reasons why you think that you should receive pocket money. Are these good enough reasons? Why? So how much pocket money do you think you should receive each week?
- Which of these two puzzles was easier? Why?

Answers

Page 1

Money Matters
2 pence = 8 shells
5 pence = 20 shells
10 pence = 40 shells
20 pence = 80 shells
50 pence = 200 shells
360s are more than 85p.
Answers will vary.

What's the Problem?
5 + 3 × 7 = 26   20 − 4 × 3 = 8
12 − (5 + 6) + 4 = 5   5 × 8 − 25 ÷ 5 = 35
Explanations will vary.

Page 2

Money Matters
Results of the investigation will vary.

At Home
Results of the investigation will vary.

The Language of Maths
$7^2 = 49$ $10^2 = 100$ $19^2 = 361$
$12^2 = 144$ $30^2 = 900$ $80^2 = 6400$

Page 3

Money Matters
Amounts, lists and reasoning will vary.

Sports Update
Malcolm spent 10 minutes on the running machine on the Monday before.

The Puzzler

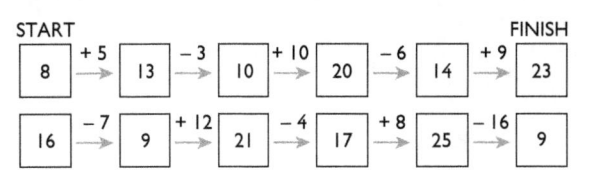

START → 8 → +5 → 13 → −3 → 10 → +10 → 20 → −6 → 14 → +9 → 23 FINISH

16 → −7 → 9 → +12 → 21 → −4 → 17 → +8 → 25 → −16 → 9

Page 4

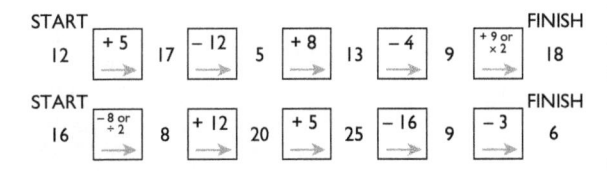

The Puzzler

START → 12 → +5 → 17 → −12 → 5 → +8 → 13 → −4 → 9 → +9 or ×2 → 18 FINISH

START → 16 → −8 or ÷2 → 8 → +12 → 20 → +5 → 25 → −16 → 9 → −3 → 6 FINISH

What's the Problem?
The bus will be full after the 11th stop.

Just before the 11th stop there will be 55 people on the bus. At the 11th stop 11 people will join the bus making a total of 66 people on the bus. So, four of the people that get on at the 11th stop will not get a seat.

Inquisitive ant

order of operations
This is the rule that states the order in which operations should be carried out in a calculation, i.e. brackets, orders (powers and square roots), division, multiplication, addition and subtraction.

Issue 14

Mixed operations

Prerequisites for learning

- Identify patterns and relationships involving numbers
- Recall and use addition and subtraction facts to 20 fluently, and derive and use related facts up to 100
- Add and subtract numbers mentally, including:
  - a two- or three-digit number and ones
  - a two- or three-digit number and tens
  - two two-digit numbers
  - adding three one-digit numbers
  - a three-digit number and hundreds
- Add and subtract numbers with up to three digits, using formal written methods of columnar addition and subtraction
- Recall and use multiplication and division facts for the 2, 3, 4, 5, 8 and 10 multiplication tables
- Write and calculate mathematical statements for multiplication and division, including for two-digit numbers by one-digit numbers, using mental and progressing to formal written methods

Resources

pencil and paper
Resource sheet 2: My notes (optional)
Resource sheet 3: Pupil self assessment booklet (optional)
selection of coins (real or play) or counters that represent £1 (optional)

Teaching support

Page 1

Technology Today

- Suggest the children draw a table to find the solution to the problem, for example:

Leo (Giles × 3)	Giles (Toby − 4)	Toby	Total
3	1	5	9
6	2	6	14
9	3	7	19
12	4	8	24
15	5	9	29
18	6	10	34
21	7	11	39

- Ensure children realise that whichever way they work out the answer to the problem, they need to keep a running total of the total number of games the three boys have, and that this amount must be between 35 and 40.

Money Matters

- Provide the children either with a selection of £1 coins (real or play) or counters that represent £1.

Issue 14 – Mixed operations

Page 2

Looking for Patterns

- Once the children have completed the grid, discuss with them the strategies they used to find the missing numbers. Did the strategies differ depending on what column the numbers were missing from?

Money Matters

- Ensure the children make, and write down, a prediction before doing any calculations.

What's the Problem?

- Children use trial and improvement to find the three numbers whose sum is 16 and product is 140.

Page 3

What's the Problem?

- Diagrams will vary to show how many people are in the Herne family. You may wish to introduce the children to a version of a family tree diagram to help them explain their answer, for example:

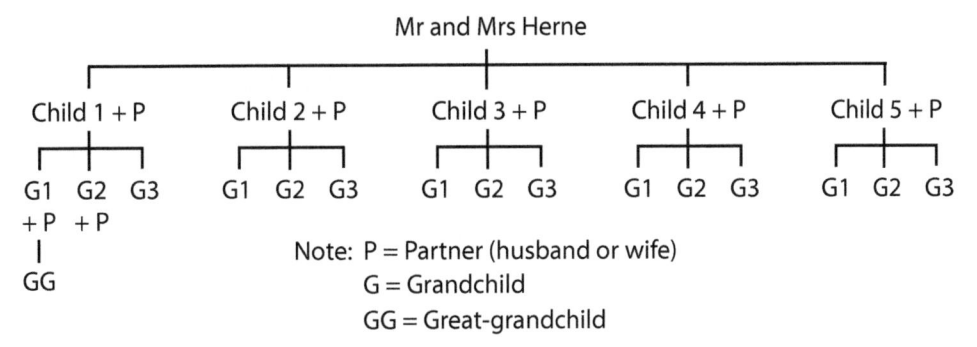

```
                            Mr and Mrs Herne
       ┌──────────┬──────────┬──────────┬──────────┐
   Child 1 + P  Child 2 + P  Child 3 + P  Child 4 + P  Child 5 + P
    ┌──┼──┐     ┌──┼──┐     ┌──┼──┐     ┌──┼──┐     ┌──┼──┐
   G1 G2 G3    G1 G2 G3    G1 G2 G3    G1 G2 G3    G1 G2 G3
   +P +P
    │
   GG
```

Note: P = Partner (husband or wife)
G = Grandchild
GG = Great-grandchild

Money Matters

- Provide the children with a selection of coins (real or play).
- Tell the children that there is more than one combination of nine coins that Stanley could have. How many different combinations can they find?

The Puzzler

- As AA × A gives a two-digit answer, A must be 1, 2 or 3.
  If A = 1, then AA × A would equal AA.
  If A = 3, then B = 9, which does not fit AA + A = AB.
  So A = 2 and B = 4.
- Tell the children that one of the values is twice the other value (see Answers).
- If AA × A = BB and AA + A = AC, what do A, B and C stand for? (Answer A = 3, B = 9 and C = 6).

Page 4

The Puzzler

- Ensure children realise that:
  – each shape represents not just a number or operator, but a mathematical operation, e.g. × 2 or + 14
  – the same shape represents the same operation, e.g. □ = × 4.

AfL

- How did you work out the answer to this problem? Did you draw a table or a diagram to help you?
- How did you work out what the square / circle number was?
- What patterns did you notice in this puzzle? How did this help you?

Issue 14 – Mixed operations

Answers

Page 1

Technology Today
Leo has 21 computer games, Giles has 7 and Toby has 11.

Money Matters
Allison has saved £24 and Michael has saved £8.
Explanations will vary.

Page 2

Looking for Patterns

□	○	□ + ○	□ × ○
5	9	14	45
8	7	15	56
3	4	7	12
7	10	17	70
9	6	15	54
12	8	20	96
6	15	21	90
14	7	21	98

Money Matters
20p a day for 365 days: £73.00; 366 days: £73.20
£1.50 a week for 52 weeks gives £78.
The £1.50 a week plan would give you the most money at the end of the year.

What's the Problem?
4, 5 and 7 years.

Page 3

What's the Problem?
There are 30 people in the Herne family.
Diagrams will vary.

Money Matters
There are many different combinations of nine coins that Stanley could have, including:

1p, 2p, 2p, 5p, 5p, 5p, 10p, 10p, 20p
or:
1p, 2p, 2p, 5p, 10p, 10p, 10p, 10p, 10p
or:
1p, 1p, 2p, 2p, 2p, 2p, 10p, 20p, 20p.

The Puzzler
A = 2, B = 4
(22 × 2 = 44 and 22 + 2 = 24)
Explanations will vary.

Page 4

The Puzzler
□ = × 4   ◇ = + 18   ○ = ÷ 2   △ = − 6

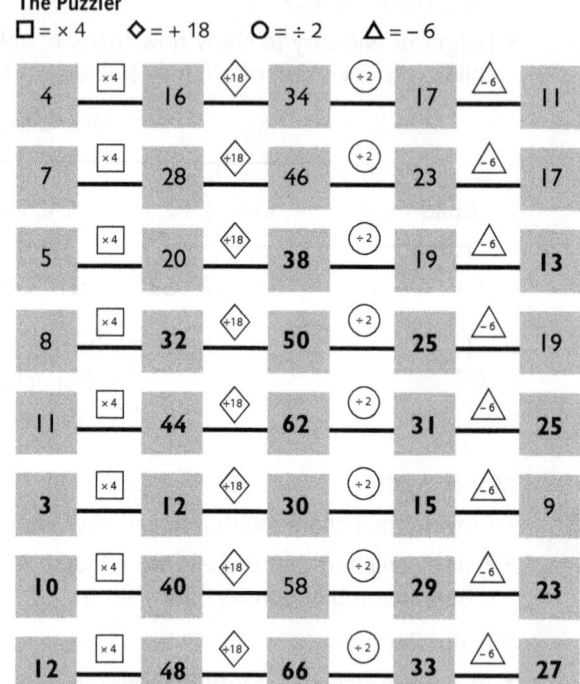

Start	×4	+18	÷2	−6
4	16	34	17	11
7	28	46	23	17
5	20	38	19	13
8	32	50	25	19
11	44	62	31	25
3	12	30	15	9
10	40	58	29	23
12	48	66	33	27

Inquisitive ant

quotient
This is the number of times one number can be divided into another. It is the answer to a division calculation.

Issue 15

Mixed operations

Prerequisites for learning

- Identify patterns and relationships involving numbers
- Recall and use addition and subtraction facts to 20 fluently, and derive and use related facts up to 100
- Add and subtract numbers mentally, including:
  - a two- or three-digit number and ones
  - a two- or three-digit number and tens
  - two two-digit numbers
  - adding three one-digit numbers
  - a three-digit number and hundreds
- Add and subtract numbers with up to three digits, using formal written methods of columnar addition and subtraction
- Recall and use multiplication and division facts for the 2, 3, 4, 5, 8 and 10 multiplication tables
- Write and calculate mathematical statements for multiplication and division, including for two-digit numbers by one-digit numbers, using mental and progressing to formal written methods

Resources

pencil and paper
Resource sheet 2: My notes (optional)
Resource sheet 3: Pupil self assessment booklet (optional)
counters, or similar (optional)
computer with Internet access

Teaching support

Page 1

The Puzzler
- Tell the children that for both puzzles, the numbers are less than 10.

Technology Today
- Children investigate the development of computer memory size and data storage over the past 10 years.

Sports Update
- Suggest children use trial and improvement, starting at around 24 points for Surinder.

Page 2

Let's Investigate
- Children investigate what happens if you choose a two-digit number greater than 50.

Let's Investigate
- Prior to carrying out this investigation, ask the children to predict which section in the table they think will have the most calculations.
- Using the digits 2, 5 and 8 the following six three-digit numbers can be made: 258, 285, 528, 582, 825 and 852.

Page 3

Focus on Science
- Encourage the children to use fractions when writing their statements comparing the number of bones in different parts of the body.

Issue 15 – Mixed operations

What's the Problem?
- Encourage the children to write down all their working out as they solve this problem, including all the calculations they performed.
- Tell the children that Lakshmi's brothers are twins.
- Once the children have solved the problem, provide an opportunity for them to talk about, and compare, their methods of working.

Page 4

What's the Problem?
- Suggest the children use a diagram to assist them in their thinking, for example:

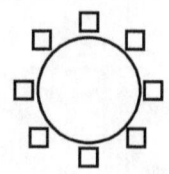

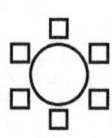

- If necessary, tell the children to draw 11 tables. Provide them with 80 counters and ask them to arrange the counters around the tables.

- Can the children use their knowledge of the multiplication facts to work out which two multiplication tables facts when added together total 80? $(4 \times 6) + (7 \times 8)$

Sports Update 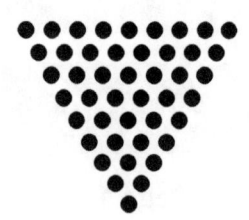 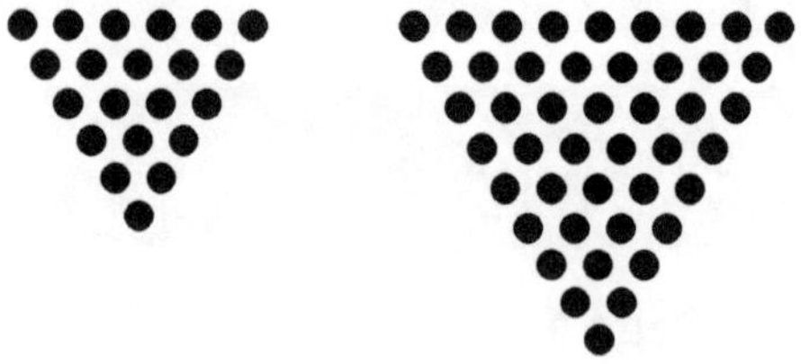
- Suggest the children draw diagrams to assist them with finding the solutions to the problems, for example:
- Ask the children to look for patterns to help them in writing a calculation for finding the total number of pins, for example:
  $1 + 2 + 3 + 4 = 10$
  $1 + 2 + 3 + 4 + 5 + 6 = 21$
  $1 + 2 + 3 + 4 + 5 + 6 + 7 + 8 + 9 = 45$
- Do the children recognise the total number of pins as being triangular numbers?

Looking for Patterns
- After two rounds of counting $\frac{1}{2} \times \frac{1}{2} = \frac{1}{4}$ are left standing, which is 6.
- After three rounds of counting $\frac{1}{2} \times \frac{1}{4} = \frac{1}{8}$ (or $\frac{1}{2} \times \frac{1}{2} \times \frac{1}{2} = \frac{1}{8}$) are left standing, which is 3.
- Provide the children with 24 counters (or similar) to represent the children in the problem.

AfL

- What strategies did you use to work out what the numbers were?
- What patterns did you notice as a result of this investigation?
- Tell me two of your most interesting statements about the bones in the human body.
- What patterns did you notice while you were doing this investigation / activity? How did these help you find the answer(s)?

Answers

Page 1

The Puzzler
4
2

Technology Today
Approximately 2700 bytes.

Sports Update
Surinder has scored 42 points.

Page 2

Let's Investigate
You get the number you started with.
This happens because:
- the number has been divided by 10 twice and multiplied by 100 once, therefore the operations cancel each other out.
- the number has been subtracted from a multiple of 10 four times, bringing it back to the number first chosen.

Let's Investigate

Answer between	Calculation
1–200	285 – 258 = 27; 582 – 528 = 54; 852 – 825 = 27
201–400	528 – 258 = 270; 528 – 285 = 243; 582 – 258 = 324; 582 – 285 = 297; 825 – 528 = 297; 825 – 582 = 243; 852 – 528 = 324; 852 – 582 = 270
401–600	825 – 258 = 567; 825 – 285 = 540; 852 – 258 = 594; 852 – 285 = 567; 285 + 258 = 543
601–800	528 + 258 = 786
801–1000	528 + 285 = 813; 582 + 258 = 840; 582 + 285 = 867
1001–1200	582 + 528 = 1110; 825 + 258 = 1083; 825 + 285 = 1110; 852 + 258 = 1110; 852 + 285 = 1137
1201–1400	825 + 528 = 1353; 852 + 528 = 1380
1401–1600	825 + 582 = 1407; 852 + 582 = 1434
1601–1800	852 + 825 = 1677

Page 3

Focus on Science
Results of the investigation will vary.

What's the Problem?
Lakshmi's brothers are two years old and her sister is four years old.

Page 4

What's the Problem?
They use four small tables for six people and seven large tables for eight people.

Sports Update
In a giant game of bowling with six rows of pins there are 21 pins altogether.
In a monster game of bowling with nine rows of pins there are 45 pins altogether.

Looking for Patterns
There are six children left standing after two counts round the circle.
There are three children left standing after three counts round the circle.

Inquisitive ant

sum
The total amount resulting when two or more numbers or quantities are added together.

Issue 16

Mixed operations

Prerequisites for learning

- Identify patterns and relationships involving numbers
- Recall and use addition and subtraction facts to 20 fluently, and derive and use related facts up to 100
- Add and subtract numbers mentally, including:
  - a two- or three-digit number and ones
  - a two- or three-digit number and tens
  - two two-digit numbers
  - adding three one-digit numbers
  - a three-digit number and hundreds
- Add and subtract numbers with up to three digits, using formal written methods of columnar addition and subtraction
- Recall and use multiplication and division facts for the 2, 3, 4, 5, 8 and 10 multiplication tables
- Write and calculate mathematical statements for multiplication and division, including for two-digit numbers by one-digit numbers, using mental and progressing to formal written methods
- Begin to understand and use the order of operations, including brackets
- Solve logic problems

Resources

pencil and paper
Resource sheet 2: My notes (optional)
Resource sheet 3: Pupil self assessment booklet (optional)
six blank cards
six counters, or similar (optional)
computer with Internet access

Teaching support

Page 1

What's the Problem?

- What if Drisse spent £3.65? What combination of the three different chocolate bars does he buy?

Let's Investigate

- The most important aspect of this investigation is not the different calculations that children write, rather it is the various ways in which the children are able to sort their calculations.
- When children have written a varied selection of calculations, arrange them into pairs and ask them to compare and sort their combined calculations. Through paired discussion, children will be more able to sort their calculations using a variety of different criteria.

The Puzzler

- How many different ways could you make a total of 99 with the fewest number of rolls?

Page 2

Let's Investigate

- Some children may need help in getting started with this investigation and in organising their work. You may wish to suggest to the children that they organise their work using a table, for example:

	\multicolumn{6}{c}{Number of houses}
	1	2	3	4	5	6
Number of letters	6	1 + 5 2 + 4 3 + 3	1 + 1 + 4 1 + 2 + 3 2 + 2 + 2	1 + 1 + 1 + 3 1 + 1 + 2 + 2	1 + 1 + 1 + 1 + 2	1 + 1 + 1 + 1 + 1 + 1

142

Issue 16 – Mixed operations

- Provide children with six blank cards (or similar) to represent the six letters and six counters (or similar) to represent the six houses.

Money Matters
- 5 × £7 = £35 and 3 × £12 = 36. No other combination will fit.
- Ensure children write about their method of working. When children have completed this activity, arrange them into pairs or groups to discuss their different methods.

The Puzzler
- Tell the children to think of a context involving measuring. If this is no clue then tell them to think about time.

Page 3

The Language of Maths
- Although these problems involve no actual mathematics, they do require the children to problem solve, reason and think logically. For these reasons you may wish to ask the children to work in pairs.

What's the Problem?
- Ben builds $\frac{1}{12}$ of a boat in a day. Noah builds $\frac{1}{6}$ of a boat in a day. Together they can build $\frac{1}{12} + \frac{1}{6} = \frac{1}{12} + \frac{2}{12} = \frac{3}{12} = \frac{1}{4}$ of a boat in a day.
- Suggest children work out what fraction of a boat each of them can build in a day. Then add the two fractions together to find how much they can build together in a day.

The Puzzler
- One way of solving this puzzle is to halve the sum and add or subtract half the difference to find the two numbers:
sum = 23
difference = 9
So 23 ÷ 2 = $11\frac{1}{2}$ and 9 ÷ 2 = $4\frac{1}{2}$
$11\frac{1}{2} + 4\frac{1}{2} = 16$
$11\frac{1}{2} - 4\frac{1}{2} = 7$
- A simpler method is using trial and improvement. Pick a number, add 9 to it and add the resulting number to the original number. If the sum is too big, subtract one from the number and try again; if the sum is too small, add one to the number and try again.

Page 4

The Puzzler
- Ask the children to write one division calculation sentence and one subtraction calculation sentence, each using A, B, C and D, where the same value for each letter fits in both calculations (D ÷ B ÷ C = A and D − B − C = A).

The Arts Roundup
- This activity will require children to use the internet to find the cost of different types of entertainment.
- It is recommended that children work in pairs for this activity so that they can share ideas and discuss their reasoning.

Let's Investigate
- Once children have completed the activity, arrange them into pairs to compare and discuss their calculations. For each number 1 to 40, children decide which calculation is the most compact and efficient.

AfL

- What strategies did you use when solving this problem / puzzle? Did you use trial and improvement or did you use a different strategy?
- Tell me three of the most interesting calculations that you wrote that have an answer of 12.
- Tell me the results of your investigation. Why have you made those conclusions?
- When writing calculations for each number to 40, what patterns did you notice? How were these able to help you find the calculations needed for creating other numbers?

Issue 16 – Mixed operations

Answers

Page 1

What's the Problem?
Drisse bought one Choco-Melto, two Choc-Blocs and two Chocolate Delights.

Let's Investigate
Calculations and criteria for sorting will vary.

The Puzzler
The fewest number of rolls needed to make a total of 99 is 4 (e.g. 60 + 30 + 6 + 3 and 50 + 40 + 6 + 3).

Page 2

Let's Investigate
There are 11 different ways that the postman could deliver all six letters to the six different houses.

Money Matters
Five children's tickets and three adult's tickets were bought.

The Puzzler
When you add two hours to 11 o'clock you get 1 o'clock.

Page 3

The Language of Maths
Henry is looking at a picture of his son.
A grandmother, mother and daughter went shopping.

What's the Problem?
It takes Ben and Noah four days to build a boat together.

The Puzzler
7 and 16 have a difference of 9 and a sum of 23.

Page 4

The Puzzler
A = 1, B = 2 and C = 3 (or different combinations, e.g. A = 3, B = 2, C = 1), D = 6.
Explanations will vary.

The Arts Roundup
Results of the investigation will vary.

Let's Investigate

1 = 2 − 1	11 = 13 − 2	21 = 24 − 3	31 = 34 − 2 − 1
2 = 4 − 2	12 = 4 × 3	22 = 23 − 1	32 = 34 − 2
3 = 2 + 1	13 = (4 × 3) + 1	23 = 24 − 1	33 = 34 − 1
4 = 3 + 1	14 = (4 × 3) + 2	24 = 4 × 3 × 2	34 = (14 + 3) × 2
5 = 3 + 2	15 = 12 + 3	25 = 24 + 1	35 = 34 + 1
6 = 2 × 3	16 = 14 + 2	26 = 24 + 2	36 = 34 + 2
7 = 4 + 3	17 = 14 + 3	27 = 24 + 3	37 = 34 + 2 + 1
8 = 4 + 3 + 1	18 = 21 − 3	28 = 24 + 3 + 1	38 = 41 − 3
9 = 4 + 3 + 2	19 = 14 + 3 + 2	29 = 31 − 2	39 = 41 − 2
10 = (2 × 3) + 4	20 = (3 + 2) × 4	30 = (13 × 2) + 4	40 = 43 − 2 − 1

Other solutions are possible.

Inquisitive ant

difference
The amount by which one quantity is greater or smaller than another.

Issue 17
Fractions

Prerequisites for learning

- Recognise that tenths arise from dividing an object into 10 equal parts and in dividing one-digit numbers or quantities by 10
- Recognise, find and write fractions of a discrete set of objects: unit fractions and non-unit fractions with small denominators
- Recognise and use fractions as numbers: unit fractions and non-unit fractions with small denominators
- Recognise and show, using diagrams, equivalent fractions with small denominators
- Compare and order unit fractions, and fractions with the same denominators
- Add and subtract fractions with the same denominator within one whole

Resources

pencil and paper
Resource sheet 2: My notes (optional)
Resource sheet 3: Pupil self assessment booklet (optional)
Resource sheet 8: 1 cm squared paper
Resource sheet 9: 2 cm squared paper
ruler
tape measure
blue, red and green coloured pencils
interlocking cubes
calculator (optional)

Teaching support

Page 1

Let's Investigate

- Remind the children that they are only expected to find an approximate fraction.
- Once the children have worked out an approximation, ensure that they spend time writing about what they did. Then hold a discussion with different pairs of children discussing their method. How could they make their approximation more accurate?

The Puzzler

- Ensure children can confidently find equivalent fractions, and also add fractions with like denominators.
- Between them, the first, second and third daughters will inherit $\frac{7}{8}$ of the kingdom, i.e. $\frac{1}{4}\left(\frac{2}{8}\right) + \frac{2}{8}\left(\frac{1}{4}\right) + \frac{3}{8}$. So the youngest daughter will inherit $\frac{1}{8}$ of the kingdom.
- Suggest the children drawn a circle divided into eighths to represent the kingdom.

Page 2

Let's Investigate

- Suggest the children use squared paper to draw their diagrams.
- For each question, what fraction of the chocolate bars does each person receive? What do you notice?

What's the Problem?

- If appropriate, discuss with the children how one-third can be thought of as one child out of every three children, and two-thirds as two children out of every three children. This will begin to introduce the children to the language of proportion.

145

Page 3

The Puzzler
- What if you changed:
  - colouring $\frac{1}{6}$ of the circles blue to $\frac{1}{8}$ of the circles blue?
  - colouring $\frac{1}{4}$ of the circles red to $\frac{1}{3}$ of the circles red?
  - colouring $\frac{1}{12}$ of the circles green to $\frac{1}{6}$ of the circles green?

The Puzzler
- Suggest that children divide the square into a 4 × 4 grid.

Page 4

Construct
- Ask the children to construct a model that has half of its cubes red, one-quarter of its cubes blue and the rest of its cubes yellow.
- Children write a description using fractions similar to that in the issue. They then ask a friend to use the description to construct the model.

What's the Problem?
- Monica sleeps for 1 hour longer each day.

  1 hour × 365 days ÷ 24 hours = 15·21 days

  or $\dfrac{1 \text{ hour} \times 365 \text{ days}}{24 \text{ hours}}$ = 15·21 days

- When writing about how they worked out the answer, encourage the children to express their working out as a fraction.
- You may wish to suggest that the children use a calculator to work out the answer to this problem.

The Language of Maths
- Encourage the children to write their statements with fractions that have been reduced to their simplest form.
- Once children have written several statements, arrange the children into pairs to compare and discuss their statements.

AfL

- How did you work out the answer to this problem? Did you use a diagram to help you?
- Explain this diagram to me. How does it show this equation / this relationship / this fraction?
- Explain to me how you worked out the answer / approximation.
- Explain your model(s) to me. How do these two different models show the same relationship?
- Tell me one of your statements. How else could you express this?

Answers

Page 1

Let's Investigate
Results of the investigation will vary.

The Puzzler
The king's youngest daughter will inherit $\frac{1}{8}$ of his kingdom.

Page 2

Let's Investigate
2 chocolate bars between 5 people:

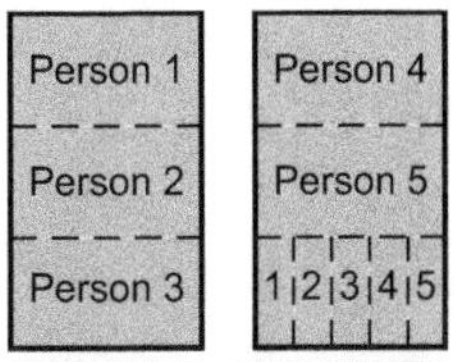

3 chocolate bars between 5 people:

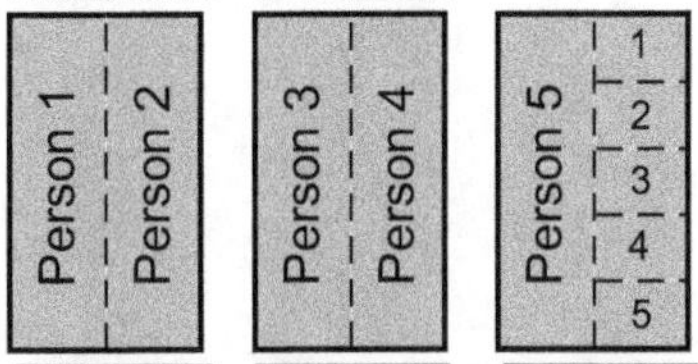

3 chocolate bars between 6 people:

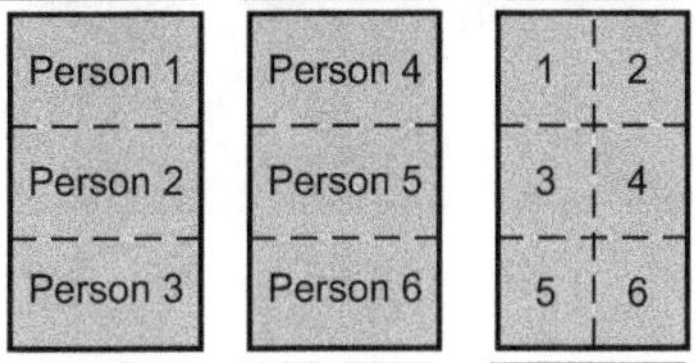

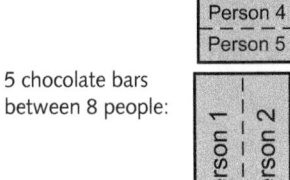

3 chocolate bars between 8 people:

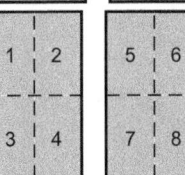

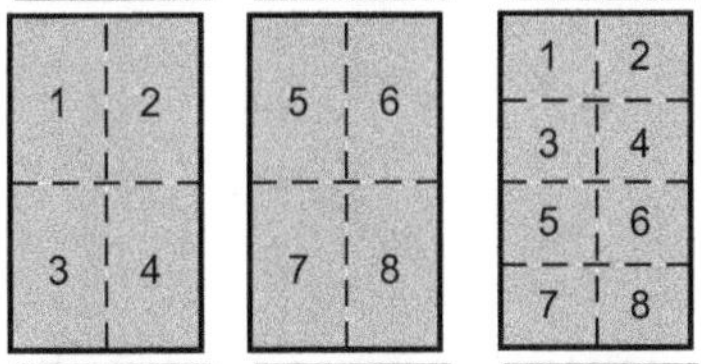

3 chocolate bars between 10 people:

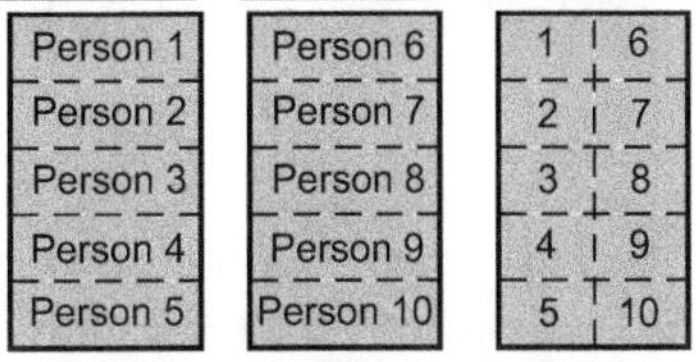

5 chocolate bars between 8 people:

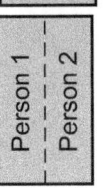

What's the Problem?
Annabelle invited 12 friends to her party.

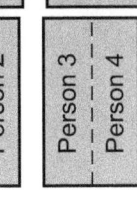

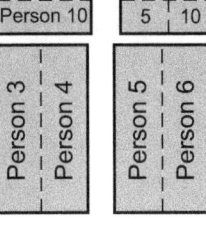

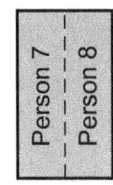

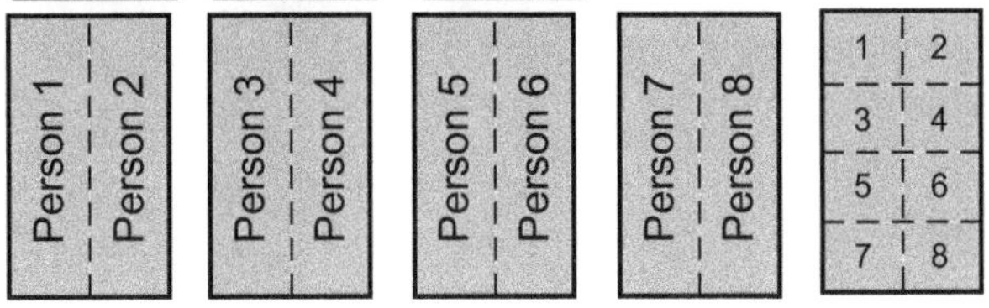

Page 3

The Puzzler
Colouring of the array will vary. However, there should be:
- 4 blue circles
- 6 red circles
- 14 circles not coloured
- $\frac{10}{24}$ ($\frac{5}{12}$) of the circles are coloured
- $\frac{14}{24}$ ($\frac{7}{12}$) of the circles are not coloured.

Then:
- 12 circles should be coloured green
- 22 circles are now coloured
- $\frac{22}{24}$ ($\frac{11}{12}$) of the circles are now coloured.

The Puzzler

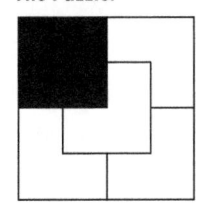

Page 4

Construct
Models will vary. However, one possible solution is for a model of 24 interlocking cubes where 12 cubes are red, 8 are blue, 3 are green and 1 is yellow.

What's the Problem?
Yes. Monica does sleep 15 days a year longer than Jamila.

The Language of Maths
Statements will vary.

Inquisitive ant

numerator and denominator
The numerator is the top number in a fraction.
The denominator is the bottom number in a fraction.

Issue 18

Fractions

Prerequisites for learning

- Recognise that tenths arise from dividing an object into 10 equal parts and in dividing one-digit numbers or quantities by 10
- Recognise, find and write fractions of a discrete set of objects: unit fractions and non-unit fractions with small denominators
- Recognise and use fractions as numbers: unit fractions and non-unit fractions with small denominators
- Recognise and show, using diagrams, equivalent fractions with small denominators
- Compare and order unit fractions, and fractions with the same denominators
- Add and subtract fractions with the same denominator within one whole
- Begin to understand decimal notation
- Relate fractions to division

Resources

pencil and paper
Resource sheet 2: My notes (optional)
Resource sheet 3: Pupil self assessment booklet (optional)
two sheets of A4 paper
ruler
large sheet of paper, i.e. A3 or A2 (optional)
calculator

Teaching support

Page 1

The Language of Maths

- This activity is designed to introduce the children to the concept of proportion and how it describes a part (or parts) of a whole and can be expressed as a fraction.

The Puzzler

- $\frac{1}{2} + \frac{1}{5} = \frac{5}{10} + \frac{2}{10} = \frac{7}{10}$, so the missing fraction needed to make 1 is $1 - \frac{7}{10} = \frac{3}{10}$.
- Ensure children can confidently find equivalent fractions, and also add fractions with like denominators.
- What if $\frac{1}{2}$ had sent $\frac{1}{5}$ off to find a pair of identical fractions to make 1? ($\frac{3}{20}$ and $\frac{3}{20}$)

Page 2

Let's Investigate

- This activity helps children to develop a feel for the relative size of fractions.
- Once children have completed this issue, you may wish to discuss with them the similarities between this activity and the first The Language of Maths activity on page 4, and how both activities use diagrams to compare fractions.
- Ask the children to use a larger sheet of paper and to draw a 40 cm, 80 cm or 100 cm line as their unit.
- Ask the children to draw vertical rather than horizontal lines.

Let's Investigate

- This activity is designed to introduce children to decimal fractions and the equivalence between fractions and decimals.
- Ensure children realise that the calculator represents the fraction in decimal form and that the decimal point separates the whole number from the fraction. Children also need to understand that as proper fractions are fractions less than 1, the whole number (the number to the left of the decimal point) is always zero.
- Encourage children to spot the recurring pattern in $\frac{1}{11}$.

Page 3

Let's Investigate

- Ensure children complete this activity before starting on the next activity (🐛 Let's Investigate).
- This activity is designed to introduce children to decimal fractions and the equivalence between decimal and fraction forms of tenths.
- Ensure children realise that the calculator represents the fraction in decimal form and that the decimal point separates the whole number from the fraction. Children also need to understand that as proper fractions are fractions less than 1, the whole number (the number to the left of the decimal point) is always zero and that for tenths (e.g. $\frac{2}{10}$) the top number in the fraction (the numerator) is written to the right of the decimal point (i.e. 0·2).

Let's Investigate

- Ensure children have completed the 🐛 Let's Investigate activity above before starting on this activity.
- This activity is designed to introduce children to decimal fractions and the equivalence between decimal and fraction forms of hundredths.
- Ensure children realise that the calculator represents the fraction in decimal form and that the decimal point separates the whole number from the fraction. Children also need to understand that as proper fractions are fractions less than 1, the whole number (the number to the left of the decimal point) is always zero and that for hundredths (e.g. $\frac{2}{100}$ or $\frac{38}{100}$) the top number in the fraction (the numerator) is written as two digits to the right of the decimal point (i.e. 0·02 or 0·38).

Page 4

The Language of Maths

- If children are unfamiliar with a fraction wall you may need to spend some time discussing it with them.
- Once children have written several statements, arrange them into pairs and ask them to discuss and compare their statements.
- Once children have completed this issue, you may wish to discuss with them the similarities between this activity and the first 🐛 Let's Investigate activity on page 2, and how both activities use diagrams to compare fractions.

The Language of Maths

- Encourage children to include a selection of both exact and approximate statements.
- Once children have written several statements, arrange them into pairs and ask them to discuss and compare their statements.

AfL

- Tell me one of your statements. How else could you express this?
- How did you work out the answer to this problem? Did you use a diagram to help you?
- Explain this diagram to me. How is it different from this diagram?
- What does this fraction mean? How else can you express this fraction?
- What are *tenths / hundredths*?
- What is this fraction as a decimal?
- What is this decimal as a fraction?
- Are these two fractions the same / equivalent? How do you know?
- Tell me a fraction less than / greater than / equivalent to …

Issue 18 — Fractions

Answers

Page 1

The Language of Maths
Statements will vary.

The Puzzler
$\frac{3}{10}$

Page 2

Let's Investigate

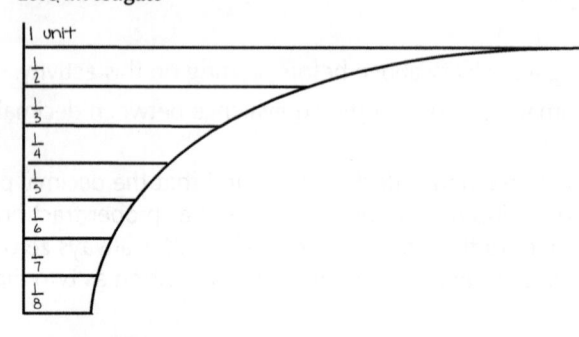

[Diagram showing a unit bar divided into fractions $\frac{1}{2}, \frac{1}{3}, \frac{1}{4}, \frac{1}{5}, \frac{1}{6}, \frac{1}{7}, \frac{1}{8}$]

Let's Investigate
$\frac{1}{2} = 0.5, \frac{1}{3} = 0.3333, \frac{1}{4} = 0.25, \frac{1}{5} = 0.2, \frac{1}{6} = 0.16666,$
$\frac{1}{7} = 0.142857, \frac{1}{8} = 0.125, \frac{1}{9} = 0.1111, \frac{1}{10} = 0.1,$
$\frac{1}{11} = 0.0909, \frac{1}{12} = 0.08333$
These numbers are referred to as decimals.
Explanations will vary.

Page 3

Let's Investigate
$\frac{1}{10} = 0.1, \frac{2}{10} = 0.2, \frac{3}{10} = 0.3, \frac{4}{10} = 0.4, \frac{5}{10} = 0.5, \frac{6}{10} = 0.6, \frac{7}{10} = 0.7,$
$\frac{8}{10} = 0.8, \frac{9}{10} = 0.9$
Explanations will vary.

Let's Investigate
$\frac{1}{100} = 0.01, \frac{2}{100} = 0.02, \ldots \frac{99}{100} = 0.99$
Explanations will vary.

Page 4

The Language of Maths
Statements will vary.

The Language of Maths
Statements will vary.

Inquisitive ant

unit fraction
A fraction with a numerator of 1, e.g. $\frac{1}{2}, \frac{1}{4}, \frac{1}{10}$.

Issue 19
Fractions

Prerequisites for learning

- Identify patterns and relationships involving numbers
- Recognise that tenths arise from dividing an object into 10 equal parts and in dividing one-digit numbers or quantities by 10
- Recognise, find and write fractions of a discrete set of objects: unit fractions and non-unit fractions with small denominators
- Recognise and use fractions as numbers: unit fractions and non-unit fractions with small denominators
- Recognise and show, using diagrams, equivalent fractions with small denominators
- Compare and order unit fractions, and fractions with the same denominators
- Add and subtract fractions with the same denominator within one whole
- Write and calculate mathematical statements for multiplication and division, including for two-digit numbers by one-digit numbers, using mental and progressing to formal written methods

Resources

pencil and paper
Resource sheet 2: My notes (optional)
Resource sheet 3: Pupil self assessment booklet (optional)
calculator
counters (or similar)

Teaching support

Page 1

Focus on Science

- Discuss with the children the link between fractions and division, i.e. $\frac{1}{5} \times 255 = 255 \div 5$.

Money Matters

- One method for working out the answer is to draw a picture, as shown here.

$\frac{1}{2} \times 24 = 12$ $12 \times 50p = £6$
$\frac{1}{3} \times 24 = 8$ $8 \times £1 = £8$
$\frac{1}{6} \times 24 = 4$ $4 \times £2 = £8$
                               $\overline{£22}$

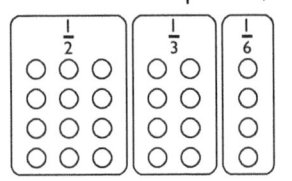

Page 2

Money Matters

- Ensure children are confident in finding both unit ($\frac{1}{2}$, $\frac{1}{10}$, $\frac{1}{20}$, $\frac{1}{5}$) and non-unit ($\frac{3}{10}$, $\frac{3}{20}$) fractions of amounts.
- Before children work out how much is spent each year on the five different categories, ensure that they have accurately worked out the total amount of money earned, i.e. £24 000.

Money Matters

- Children choose different combinations of three coins and investigate what fractions of the total they can make where the result is a whole amount.

Page 3

Money Matters

- Ensure children realise that when they make their own "Would you rather …?" problems that they need to have the answers to the two calculations in each problem as close to each other as possible.

What's the Problem?

- Conroy ate $24 \times \frac{1}{3} = 8$ biscuits, leaving 16.
  Daniel ate $16 \times \frac{1}{2} = 8$ biscuits, leaving 8.

What's the Problem?

- $24 \times \frac{3}{4} = 18$ children liked lemonade, $24 \times \frac{2}{3} = 16$ children liked orange juice, and $24 \times \frac{1}{2} = 12$ children liked both.
- Therefore $18 - 12 = 6$ children liked only lemonade and $16 - 12 = 4$ children liked only orange juice. So, $12 + 6 + 4 = 22$ children liked lemonade, orange juice or both. Which leaves two children ($24 - 22 = 2$) who liked neither lemonade nor orange juice.
- Use 24 counters (or similar) to represent the children.

Page 4

What's the Problem?

- Suggest the children use trial and improvement to find the solution to the problem.
- Use 32 counters (or similar) to represent the sweets.

The Puzzler

- Ensure children understand the terms "numerator" and "denominator". You may also need to remind the children of equivalent fractions.
- When you add 2 to both the numerator and denominator of $\frac{1}{4}$ you get $\frac{1+2}{4+2} = \frac{3}{6}$ or $\frac{1}{2}$, which is double $\frac{1}{4}$.
- Suggest the children start by using different unit fractions, e.g. $\frac{1}{2}, \frac{1}{3}, \frac{1}{4}$.
- You may also wish to ask children to work in pairs to solve this puzzle.

What's the Problem?

- Towser eats $\frac{2}{3} \times 2 = 1\frac{1}{3}$ tins of food a day, or four tins in three days. So 12 tins will last $\frac{12}{4} \times 3 = 9$ days.

AfL

- How did you work out the answers to the diamonds activity? Show me how you worked out one of the answers using the calculator. What does this number mean?
- What did you find out in this investigation?
- Which amount of money would you rather have? Why?
- Tell me how you worked out the answer to this problem. What strategies did you use to find the answer?

Answers

Page 1

Focus on Science
The Star of the South: 51 g
The Regent: 82 g
The Kimberley Octahedron: 123 g
Millennium Star: 155 g
Star of Sierra Leone: 194 g
Cullinan Diamond: 621 g

Money Matters
Alpesh has £22 in his money box.

Page 2

Money Matters
You would earn £480 if you worked 40 hours a week.
You would earn £24 000 in a year if you worked for 50 weeks.
Money spent on:
- accommodation: £12 000
- food: £3600
- holidays: £2400
- entertainment: £1200.

You save £4800.

Money Matters
Results of the investigation will vary depending on the coins chosen.

Page 3

Money Matters
$\frac{1}{4} \times £18$ (£4.50) $\frac{1}{5} \times £29$ (£5.80) $\frac{2}{3} \times £42$ (£28) $\frac{5}{6} \times £15$ (£12.50)

What's the Problem?
Eight biscuits are left in the packet.

What's the Problem?
Two children said that they liked neither lemonade nor orange juice.

Page 4

What's the Problem?
Ellie ate 8 sweets and Josh ate 24 sweets.

The Puzzler
$\frac{1}{4}$

What's the Problem?
A box of dog food lasts Towser 9 days.

Inquisitive ant

non unit fraction
A fraction with a numerator other than one, e.g. $\frac{2}{3}, \frac{3}{4}, \frac{2}{5}$.

Issue 20

Fractions

Prerequisites for learning

- Recognise that tenths arise from dividing an object into 10 equal parts and in dividing one-digit numbers or quantities by 10
- Recognise, find and write fractions of a discrete set of objects: unit fractions and non-unit fractions with small denominators
- Recognise and use fractions as numbers: unit fractions and non-unit fractions with small denominators
- Recognise and show, using diagrams, equivalent fractions with small denominators
- Compare and order unit fractions, and fractions with the same denominators
- Add and subtract fractions with the same denominator within one whole
- Begin to recognise fractions greater than 1
- Write and calculate mathematical statements for multiplication and division, including for two-digit numbers by one-digit numbers, using mental and progressing to formal written methods

Resources

pencil and paper
Resource sheet 2: My notes (optional)
Resource sheet 3: Pupil self assessment booklet (optional)
Resource sheet 8: 1 cm squared paper
ruler
blue, red, yellow and green coloured pencils
calculator (optional)
set of 0–9 digit cards

Teaching support

Page 1

What's the Problem?

- Each tin of dog food costs £14.40 ÷ 12 = £1.20.
- So the cost of dog food for the year is $1\frac{1}{4}$ tins × 365 days × £1.20 = £547.50.
- You may wish to suggest that the children use a calculator to work out the answer to this problem.

Let's Investigate

- Ensure children have an understanding of equivalent fractions and improper fractions.
- You may need to provide the children with one or two completed number sentences to get them started with this investigation.

Page 2

What's the Problem?

- Ensure children have an understanding of equivalent fractions and how to add fractions with like denominators.
- Tamsin's furniture takes up $\frac{1}{3} + \frac{1}{12} = \frac{4}{12} + \frac{1}{12} = \frac{5}{12}$ of the floor space. So, $1 - \frac{5}{12} = \frac{7}{12}$ of the floor space is left free.

Let's Investigate

- Encourage the children to be systematic in their recording of the different calculations possible, and to look out for patterns which will help them identify other possible calculations.

Page 3

Around the World
- Canadian side: $\frac{4}{5} \times 750\,000 = 600\,000$ gallons per second
- US side: $\frac{1}{5} \times 750\,000 = 150\,000$ gallons per second
- Even though this activity requires children to find both a unit and a non-unit fraction of a large amount, given that that number is a multiple of 5, children should be encouraged to work out the answers using mental and / or written methods.

In the Past
- The number of fighting men in a Roman legion was $81 \times 6 \times 10 = 4860$ men.
- The 140 extra men (5000 − 4860) in a legion consisted of doctors, engineers and other workers.
- Two methods of calculating what fraction were centurions are:
  - Shorter method:
    80 men were commanded by 1 centurion, therefore 1 in 81 men were centurions or $\frac{1}{81}$.
  - Longer method:
    Number of centurions in a legion = $1 \times 6 \times 10 = 60$.
    Fraction of men who were centurions = (number of centurions) ÷ (number of fighting men in legion)
    $= \frac{60}{4860}$
    $= \frac{1}{81}$

Page 4

Construct
- Suggest children work out how many squares of each colour there must be on their flag before they start to design them (white: 80, blue: 60, red: 48, yellow: 20, green: 32).

What's the Problem?
- Leroy is only $\frac{4}{5}$ of Lorray's height, so $\frac{1}{5}$ of Lorray's height is 27 cm. Lorray is therefore $5 \times 27 = 135$ cm tall and Leroy is $\frac{4}{5} \times 135 = 108$ cm.
- Draw a picture to scale of Leroy and Lorray. Show that Leroy is $\frac{4}{5}$ in height and Lorray is $\frac{5}{5}$ in height. The difference in height is 27 cm, so Leroy must be 4×27 cm = 108 cm tall and Lorray must be 5×27 cm = 135 cm tall.

What's the Problem?
- Paul fills the watering can with $\frac{3}{4} \times 8 = 6$ litres of water. He puts $\frac{1}{3} \times 6 = 2$ litres on each pot.

AfL

- What else does $1\frac{1}{3}$ mean? What about $1\frac{1}{2}$ or $1\frac{1}{4}$? How else could you express $1\frac{1}{2}$ and $1\frac{1}{4}$?
- What calculations did you have to do in order to work out the answer to this problem?
- Tell me some of the fraction calculations you made. What patterns do you notice?
- Talk me though your flag design. What fraction of your flag is green?

Issue 20 – Fractions

Answers

Page 1

What's the Problem?
Tim spends £547.50 on dog food in a year.

Let's Investigate

$\frac{2}{3} - \frac{4}{6} = \frac{2}{3} - \frac{2}{3} = 0$ 　　 $\frac{3}{4} - \frac{6}{8} = \frac{3}{4} - \frac{3}{4} = 0$

$\frac{2}{3} - \frac{6}{9} = \frac{2}{3} - \frac{2}{3} = 0$ 　　 $\frac{3}{9} - \frac{2}{6} = \frac{1}{3} - \frac{1}{3} = 0$

$\frac{2}{4} - \frac{3}{6} = \frac{1}{2} - \frac{1}{2} = 0$ 　　 $\frac{4}{8} - \frac{3}{6} = \frac{1}{2} - \frac{1}{2} = 0$

Page 2

What's the Problem?

$\frac{7}{12}$

Let's Investigate

$\frac{1}{3} \times 24 = 8$ 　　 $\frac{1}{7} \times 28 = 4$ 　　 $\frac{1}{9} \times 27 = 3$

$\frac{1}{3} \times 27 = 9$ 　　 $\frac{1}{7} \times 42 = 6$ 　　 $\frac{1}{9} \times 36 = 4$

　　　　　　　　 $\frac{1}{7} \times 56 = 8$ 　　 $\frac{1}{9} \times 54 = 6$

$\frac{1}{4} \times 28 = 7$ 　　 $\frac{1}{7} \times 63 = 9$ 　　 $\frac{1}{9} \times 63 = 7$

$\frac{1}{4} \times 32 = 8$ 　　　　　　　　 $\frac{1}{9} \times 72 = 8$

$\frac{1}{4} \times 36 = 9$ 　　 $\frac{1}{8} \times 24 = 3$

　　　　　　　　 $\frac{1}{8} \times 32 = 4$

$\frac{1}{6} \times 42 = 7$ 　　 $\frac{1}{8} \times 56 = 7$

$\frac{1}{6} \times 54 = 9$ 　　 $\frac{1}{8} \times 72 = 9$

Page 3

Around the World
600 000 gallons of water go over the Canadian side of the Falls each second.
150 000 gallons of water go over the US side of the Falls each second.

In the Past
There were 4860 fighting men in a legion. $\frac{1}{81}$ were centurions.

Page 4

Construct
Flag designs will vary.
$\frac{2}{15}$ of the flag should be green.

What's the Problem?
Lorray is 135 cm tall, and Leroy is 108 cm tall.

What's the Problem?
Paul puts 2 litres of water on each pot.

Inquisitive ant

gallon
An imperial unit of measuring capacity. One gallon is approximately 4·5 litres.

Issue 21

Length

Prerequisites for learning

- Know the relationships between kilometres and metres, metres and centimetres
- Choose and use appropriate units to estimate, measure and record measurements
- Measure and draw to a suitable degree of accuracy
- Calculate mentally with whole numbers
- Develop and use written methods to add, subtract, multiply and divide whole numbers
- Make estimations and approximations

Resources

pencil and paper
Resource sheet 2: My notes (optional)
Resource sheet 3: Pupil self assessment booklet (optional)
art paper and colouring materials
collection of clothing manufactured in different countries with labels showing country of production (optional)
tape measure
computer with Internet access

Teaching support

Page 1

The Puzzler

- Suggest the children draw a scale drawing of the pond and island and use two small pieces of paper with the plank lengths marked on their edges to find a way to get across.

The Arts Roundup

- This activity is aimed at introducing to the children the idea of scale through a highly practical activity.
- You may wish to discuss with the children the accuracy of scale drawings when using direct proportion, and how this compares with the inaccuracy of scale drawings that use perspective.

Page 2

Around the World

- You may wish to provide the children with items of clothing manufactured in different countries, with labels showing country of production.
- Alternatively, ask the children to find pieces of clothing at home and comment on their country of production. They can then complete the second part of the activity back in school, i.e. finding out approximately how far each piece of clothing has travelled.
- If the children do find pieces of clothing at home, provide an opportunity for pairs or groups of children to pool their lists before working on the second part of the activity.

What's the Problem?

- Ask the children to draw a diagram representing the problem, for example:

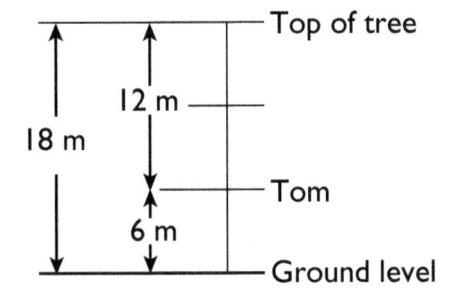

```
                ─── Top of tree
        ↑   ↑
       12 m ─┼─
  18 m  ↓
        ↑   ─── Tom
       6 m
        ↓   ─── Ground level
```

What's the Problem?

- As children are only expected to approximate the total length the class would stretch if they laid down head to foot, discuss with the children that all they really need to do is to find the "average" height of all the children in the class, and then multiply this by the number of children in the class.

Issue 21 – Length

- Discuss how this method would also work to find out the total length that the whole school would stretch. However, a more accurate measurement can be gained from finding out the average height for different year groups / age ranges, e.g. foundation stage, Years 1 and 2, Years 3 and 4, and Years 5 and 6.

Page 3

Focus on Science
- Ensure children are familiar with an eyesight testing board.
- As children construct their board ensure that the number of lines and the size of fonts they include are reasonable.
- The experiment may need to be conducted in a hall or other suitable space.

At Home
- The most important aspect of this activity is the reasons that children are able to give as to why road speed signs vary at specific locations.
- Once the children have completed the activity, ensure that there is an opportunity in class for pairs or groups of children to discuss their results.

Let's Investigate
- Children will only be able to arrive at a very rough estimate. What is important in this activity is not the answer but rather the children's reasoning and justifications for their estimate. Therefore, it is recommended that children work in pairs on this activity so that they can share ideas and discuss their reasoning.

Page 4

Around the World
- Ensure children are familiar with the terms "length" and "width".
- If appropriate, you may wish to use this activity to introduce the children to the concept of perimeter.

Sports Update
- Suggest the children draw a table to identify the pattern and work out the answer, for example:

Sam	3 km	6 km	9 km	12 km	15 km
Mats	7 km	14 km	21 km	28 km	35 km

Let's Investigate
- Before working independently on this activity, ensure the children understand that a kilometre (km) is a measure of distance.
- A pace is measured as two steps – starting from the left foot, then the right foot, then the left foot.
- This activity requires the children to make an estimate. Suggest the children inform their estimate by measuring the number of paces they take in 1 m and then scale up to estimate how many paces it would take them to walk 10 m.
- Children can then compare their estimate with the exact number.

AfL

- Describe for me the relative height of some of the things you have drawn in your picture.
- How did you go about finding out how far each piece of clothing has travelled?
- How did you arrive at your estimate / approximation? What assumptions have you made? How accurate do you think your estimation / approximation is?
- Explain to me the maths you had to do in order to work out the answer to this problem.

Answers

Page 1

The Puzzler

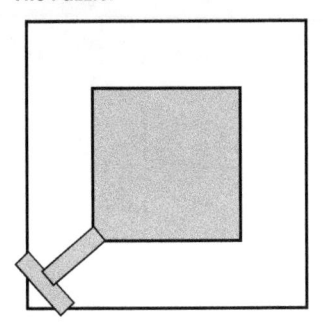

The Arts Roundup
No answers required.

Page 2

Around the World
Results of the investigation will vary.

What's the Problem?
18 metres

What's the Problem?
Results of the investigation will vary.

Page 3

Focus on Science
Results of the experiment will vary.

At Home
Results of the investigation will vary.

Let's Investigate
Answers and explanations will vary.

Page 4

Around the World
The length of the fence is 12 m and the width is 5 m.

Sports Update
Sam will have run 15 km.

Let's Investigate
Results of the investigation will vary.

Inquisitive ant

rate
The change in one quantity compared to the change in another quantity.

Issue 22
Mass

Prerequisites for learning

- Know the relationship between kilograms and grams
- Choose and use appropriate units to estimate, measure and record measurements
- Read, to the nearest division and half-division, scales that are numbered or partially numbered
- Calculate mentally with whole numbers
- Develop and use written methods to add, subtract, multiply and divide whole numbers
- Make estimations and approximations

Resources

pencil and paper
Resource sheet 2: My notes (optional)
Resource sheet 3: Pupil self assessment booklet (optional)
collection of different sized weights, including 1 kg, 500 g and 100 g weights
50 g weight (optional)
modelling clay
weighing scales
scale balance
marbles, lead beads or similar (optional)
Compare Bears or similar (optional)
calculator (optional)
counters, interlocking cubes, beads, crayons, pencils, dice, etc.

Teaching support

Page 1

Money Matters

- Suggest the children draw a table to help them organise their work, for example:

6 kg bags at £6.40 each	10 kg bags at £7.50 each	Total mass	Total cost
0	5	50 kg	£37.50
1	4	46 kg	£36.40
2	3	42 kg	£35.30
4	2	44 kg	£40.60

Construct

- This is a highly practical hands-on activity which is designed to help children get a "feel" for what 500 g and 100 g are.
- Children make 250 g, 150 g and 50 g approximate weights.

Page 2

Focus on Science

- Ensure children spend time making a prediction as to which they think is their strongest finger. Why do they think this?
- The most important aspect of this activity is the design of the test. Encourage the children to spend time thinking about the following:
   - What resources they will need
   - How they will conduct the experiment
   - How they will ensure that it is a "fair test"
   - How they will keep an ongoing record of their findings
   - Health and safety issues

What's the Problem?
- Provide the children with a scale balance, a 1 kg weight, some lead beads / marbles and additional weights that total another 1 kg (to represent the 1 kg of beans).

Page 3
Looking for Patterns
- Allow the children to use a calculator.
- Tell the children that for the last question, Joan buys a total of 13 pieces of fruit.

What's the Problem?
- Suggest the children use practical apparatus such as Compare Bears or similar.

What's the Problem?
- Towser's mass is 9 kg + $\frac{2}{3}$ more. $\frac{1}{3} + \frac{2}{3}$ = the total mass, so 9 kg = $\frac{1}{3}$ of the mass.

Page 4
Let's Investigate
- Encourage the children to think about why some children's handfuls weigh more than others.
- Tell each child to order each of the objects by weight. Which objects weigh more or less? Does each child have the same order?

What's the Problem?
- The average mass of an HB pencil is 10 g. Therefore 6000 pencils have a mass of 60 kg.
- Approximately how many trips will Ms Spry need to make before all 6000 pencils are in the stock cupboard?

AfL

- Which combination of bags of oranges offers the best value for money? How do you know?
- How did you try to make sure that your weight was as close to 500 g as possible?
- Talk me through your experiment. Tell me your results. How can you be sure that your results are valid?
- Tell me how you solved this problem / puzzle. Would you work it out the same way if you had to do this activity again? What would you do differently?
- Tell me about the predictions you made. What assumptions were your predictions based on? How accurate were they?

Answers

Page 1

Money Matters
Two 6 kg bags and three 10 kg bags give 42 kg exactly at a total cost of £35.30. This combination offers the best value for money.

Construct
Modelling clay weights will vary.

Page 2

Focus on Science
Results of the investigation will vary.

What's the Problem?
The customer and the grocer put the 1 kg weight on one side of the scales, and balance it with lead beads on the other side. They then remove the 1 kg weight and replace it with enough beans to balance the scales.

Page 3

Looking for Patterns
- One orange = 250 g
  One apple = 125 g
  One pear = 200 g
- The total mass of one orange, one apple and one pear is 575 g.
- The total mass of two oranges, one apple and two pears is 1025 g.
- Three oranges have the same mass as six apples.
- Joan buys four oranges (1 kg), four apples (500 g) and five pears (1 kg).

What's the Problem?
Four small bags of flour have the same mass as one large bag of flour.

What's the Problem?
Towser has a mass of 27 kg.

Page 4

Let's Investigate
Estimates and masses will vary.

What's the Problem?
Ms Spry would not be able to carry all 6000 pencils to the stock cupboard in one journey.

Inquisitive ant

approximation
A value that is not completely accurate or exact, but is only slightly more or less than the correct value and is therefore "near enough".

Issue 23

Capacity and volume

Prerequisites for learning

- Know the relationship between litres and millilitres
- Choose and use appropriate units to estimate, measure and record measurements
- Read, to the nearest division and half-division, scales that are numbered or partially numbered
- Calculate mentally with whole numbers
- Develop and use written methods to add, subtract, multiply and divide whole numbers
- Make estimations and approximations

Resources

pencil and paper
Resource sheet 2: My notes (optional)
Resource sheet 3: Pupil self assessment booklet (optional)
3-litre and 4-litre jugs (optional)
access to water
1-litre jug
weighing scales
two sheets of A4 paper
sticky tape
small counters or cubes
rice, pasta or sand (optional)
computer with Internet access

Teaching support

Page 1

At Home

- Once the children have completed the activity, ensure that there is an opportunity in class for pairs or groups of children to discuss their findings.

The Puzzler

- Tell the children that a glass can be moved without changing its position in the order.

Page 2

What's the Problem?

- Provide the children with a 3-litre and a 4-litre jug.

The Puzzler

- Discuss with the children how a 3-D shape, such as a cuboid, has three dimensions, and that changing any one of these dimensions changes how much the 3-D shape will hold.
- The diagram on the right illustrates why, if you double all the dimensions of a jug of water, the jug will hold eight times as much.
- Tell children to imagine a cuboid shaped jug.

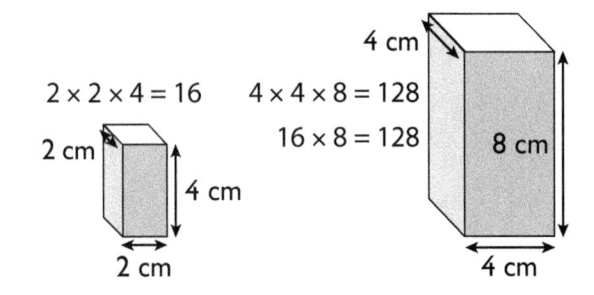

$2 \times 2 \times 4 = 16$

$4 \times 4 \times 8 = 128$

$16 \times 8 = 128$

Focus on Science

- Children may need to undertake some of this investigation at home as well as on the Internet. If so, ensure that there is an opportunity in class for pairs or groups of children to discuss their findings.

Page 3

Let's Investigate

- Ensure children realise that they are measuring the mass of the water alone, and not the combined mass of the water and the jug / container. If necessary, discuss with the children the need to first find the mass of the jug / container without water in it, and then to subtract this mass from the mass of the jug / container and water, to find the mass of the water alone.

163

At Home

- Once the children have completed the activity, ensure that there is an opportunity in class for pairs or groups of children to discuss their results.
- Can the children give reasons as to why certain types of liquids come in specific sized containers, e.g. perfume, medicines, washing liquid, soft drinks, …?

Page 4

Focus on Science

- Before setting children to work independently on this activity, ensure that they fully understand what is expected of them. Also remind the children that before they carry out their experiment, they must make a prediction as to which cylinder they think will hold the most.
- To gain a more accurate measurement, rather than using small counters or cubes, the children could use rice, pasta or sand to fill the cylinders. They will then need some measuring device to find out how much rice / pasta / sand filled the cylinders.

Let's Investigate

- To complete this investigation successfully children need to know / calculate the following:
    - The number of litres of drink a normal bottle holds, e.g. 2 litres
    - The number of glasses you can get from one bottle, e.g. approximately eight (2 litres ÷ 250 ml)
    - The number of drinks needed altogether, e.g. approximately 36 (12 × 3)
- Given the above information, five bottles of drink would need to be bought (36 ÷ 8 = $4\frac{1}{2}$).

AfL

- Tell me the results of your investigation.
- What strategies did you use to solve this problem / puzzle?
- Which list has the most / least containers? Why do you think this is?
- Were the results as you had imagined? What was the same? What surprised you? Why?

Answers

Page 1

At Home
Answers will vary.

The Puzzler
Pour the water from the second glass into the fifth glass.

Page 2

What's the Problem?
The farmer could measure out the milk by filling the 3-litre jug and pouring it into the 4-litre jug. Then fill the 3-litre jug again and use it fill up the rest of the 4-litre jug. There are now two litres left in the 3-litre jug.

The Puzzler
Eight times as much.

Focus on Science
A dishwasher uses less water than washing up by hand.

Page 3

Let's Investigate
One litre of water has a mass of 1 kg.
Results of the investigation will vary.

At Home
Results of the investigation will vary.

Page 4

Focus on Science
The short fat cylinder will hold more than the tall thin cylinder.

Let's Investigate
Results of the investigation will vary.

Inquisitive ant

justify
To give a reason or explanation for something.

Issue 24
Time

Prerequisites for learning

- Use units of time (seconds, minutes, hours, days) and know the relationships between them
- Use a calendar and know the number of days in each month, year and leap year
- Use 12-hour analogue and digital clocks to tell and write the time
- Use a.m. and p.m. notation
- Calculate time intervals and find start or end times for a given time interval
- Calculate mentally with whole numbers
- Develop and use written methods to add, subtract, multiply and divide whole numbers
- Make estimations and approximations

Resources

pencil and paper
Resource sheet 2: My notes (optional)
Resource sheet 3: Pupil self assessment booklet (optional)
computer with Internet access

Teaching support

Page 1

The Language of Maths

- Discuss with the children that while 12 is an important number in relation to time, this is not the case for other measures such as length, mass and capacity. Which number is important to these measures? (10) Can the children give reasons as to why this is?

Sports Update

- Suggest the children use a table to organise the schedule, for example:

	Start time	Finish time
Game 1		
Game 2		
Game 3		
Game 4		
Game 5		
Game 6		
Game 7		
Game 8		

Page 2

Around the World

- You may need to suggest a suitable venue for the children to use as the site for their educational visit. It might be a venue that a class have visited previously and where there may be information on record in school about the visit that the children can use.
- Remind the children that the most important aspect of this activity is the construction of the timetable. Encourage them to be as accurate as possible, accounting for all the day's events.

Looking for Patterns

- The clock strikes a total of 78 times in 12 hours:
  $1 + 2 + 3 + 4 + 5 + 6 + 7 + 8 + 9 + 10 + 11 + 12$

- In one full day, i.e. 24 hours, the clock strikes 78×2 chimes: a total of 156 times.

The Puzzler

- Suggest the children draw a diagram to assist them in working out the answer, for example:

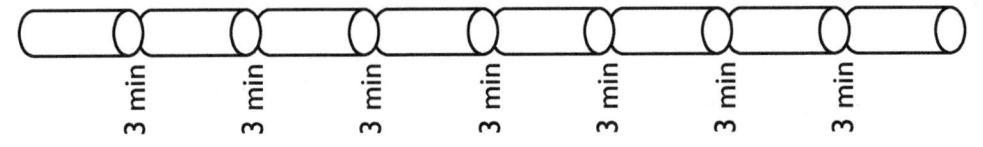

[Diagram: chain of 7 segments, each labelled "3 min"]

7×3 min = 21 min

Page 3

What's the Problem?

- Suggest the children draw a timeline and work backwards through the problem.

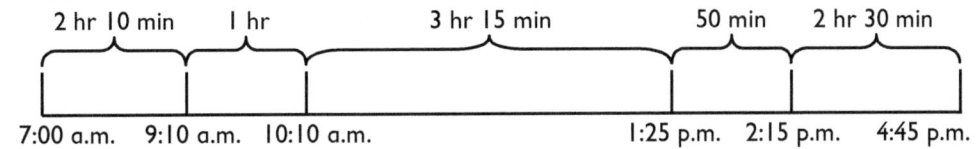

[Timeline diagram]
| 2 hr 10 min | 1 hr | 3 hr 15 min | 50 min | 2 hr 30 min |

7:00 a.m.   9:10 a.m.   10:10 a.m.   1:25 p.m.   2:15 p.m.   4:45 p.m.

- Discuss with the children that the timing and sequence of the day's events is not important, for example, the time at which Michael had lunch. Rather it is finding the total amount of time Michael spent doing various tasks throughout the day, knowing what time his work day ended, in order to be able to find the time at which Michael started work.

What's the Problem?

- Ask the children to work out which rate is cheaper: £8 for 6 hours or £6 for 4 hours? (£8 for 6 hours = £1.33 per hour / £6 for 4 hours = £1.50 per hour.) Tell the children to include all their working.

Page 4

What's the Problem?

- Ask the children to write about how they worked out the answer to the problem.

Focus on Science

- How many months have only 29 days in them? (one, February, every four years)

Around the World

- The ferry leaving Dover passes three ferries coming from Calais. It passes:
  - the first ferry after $\frac{1}{2}$ hour (this is the ferry that was already half way across when it left)
  - the second ferry after 1 hour (this is the ferry that left at the same time as it did)
  - the third ferry after $1\frac{1}{2}$ hours (the ferry that left an hour after it did).
- Suggest children draw a diagram of the ferries' positions at the time the Dover ferry leaves and every $\frac{1}{2}$ hour after that.

AfL

- What do almost all of the numbers that are concerned with time have in common? What are some of the exceptions? Why is this?
- Talk me through your proposed schedule / timetable. What assumptions have you made?
- What patterns helped you find the solution to this puzzle / answer to this problem?
- Did you draw a diagram or a table to help you answer this problem? What did you do? How did it help?

Issue 24 – Time

Answers

Page 1

The Language of Maths
Numbers will vary. However, may include the following:
- 12 months in a year
- 24 hours in a day, 60 minutes in an hour, 60 seconds in a minute
- 7 days in a week; 28, 29, 30 or 31 days in a month; 52 weeks in a year; 365 days in a year; half and quarter hours

Sports Update
Answers will vary. However, a possible schedule may be:

	Start time	Finish time
Game 1	9:00 a.m.	9:30 a.m.
Game 2	9:40 a.m.	10:10 a.m.
Game 3	10:20 a.m.	10:50 a.m.
Game 4	11:00 a.m.	11:30 a.m.
Game 5	11:40 a.m.	12:10 p.m.
Game 6	12:20 p.m.	12:10 p.m.
Game 7	1:00 p.m.	1:30 p.m.
Game 8	1:40 p.m.	2:10 p.m.

Page 2

Around the World
Timetables will vary.

Looking for Patterns
The clock strikes a total of 156 times.

The Puzzler
It takes Simon 21 minutes to join together eight pipes.

Page 3

What's the Problem?
Michael started work at 7:00 a.m.

What's the Problem?
Helen was three hours late in returning the bike.

Page 4

What's the Problem?
Oliver is 8 years old and his father is 40 years old.

Focus on Science
Seven months have 31 days. All the months of the year except for February, April, June, September and November.

Around the World
The ferry leaving Dover passes three ferries coming from Calais.

Inquisitive ant

a.m. and p.m.
a.m. (ante meridiem) is the period between midnight and noon.
p.m. (post meridiem) is the period between noon and midnight.

Issue 25

Measurement

Prerequisites for learning

- Identify patterns and relationships involving numbers
- Calculate mentally with whole numbers
- Develop and use written methods to add, subtract, multiply and divide whole numbers
- Make estimations and approximations
- Use a calendar and know the number of days in each month, year and leap year
- Use a.m. and p.m. notation

Resources

pencil and paper
Resource sheet 2: My notes (optional)
Resource sheet 3: Pupil self assessment booklet (optional)
Resource sheet 8: 1 cm squared paper
ruler
several sheets of A4 paper (preferably used sheets)
pack of playing cards
metre rule, tape measure and trundle wheel (optional)
sheet of A4 paper
selection of different envelopes
selection of different newspapers
interlocking cubes and counters (optional)

Teaching support

Page 1

What's the Problem?

- Trevor's grandfather was 60 when Trevor was born, so he is now 65 and Trevor is 5 years old.

Let's Investigate

- What if the children chose a 3 × 3 square? What patterns do they notice about the sums and differences of numbers contained in the square? Can they make any generalisations?

Page 2

Let's Investigate

- This activity, the next Let's Investigate activity and the Looking for Patterns and Let's Investigate activities on page 3, all informally introduce children to the concept of area through practical, discovery-based learning activities. You may wish to take this opportunity to introduce this concept to the children more formally once they have completed the issue. Note that the Inquisitive ant word for this issue is 'area'.
- For the first part of this activity, children should use as many A4 sheets of paper as is necessary to cover the desk.
- However, for the second part of the activity, only provide each child with enough playing cards to cover one sheet of A4 paper (approximately 11). This is so children use and apply the knowledge they acquired when using A4 paper to work out how many playing cards would be needed to cover their desk.

Let's Investigate

- This activity, the previous Let's Investigate activity and the Looking for Patterns and Let's Investigate activities on page 3, all informally introduce children to the concept of area through practical, discovery-based learning activities. You may wish to take this opportunity to introduce this concept to the children more formally once they have completed the issue. Note that the Inquisitive ant word for this issue is 'area'.
- If children can't agree on the order, they will need to take measurements of the appropriate rooms. Children can either use a metre rule, tape measure and / or trundle wheel to take these measurements.
- If it is not practicable for children to measure certain rooms where the order is disputed, decide by consensus.

Issue 25 – Measurement

Page 3

Looking for Patterns

- This activity, the next 🐜 Let's Investigate activity and the two 🐜 Let's Investigate activities on page 2, all informally introduce children to the concept of area through practical, discovery-based learning activities. You may wish to take this opportunity to introduce this concept to the children more formally once they have completed the issue. Note that the Inquisitive ant word for this issue is 'area'.
- If the children have previously completed Issue 26 or already have some understanding of the concept of perimeter, then ask the children questions similar to the following:
    - What is the perimeter of each of your rectangles?
    - Which rectangle has the smallest perimeter?
    - Which rectangle has the largest perimeter?

Let's Investigate

- This activity, the previous 🐜 Looking for Patterns activity and the two 🐜 Let's Investigate activities on page 2, all informally introduce children to the concept of area through practical, discovery-based learning activities. You may wish to take this opportunity to introduce this concept to the children more formally once they have completed the issue. Note that the Inquisitive ant word for this issue is 'area'.
- Work with the children to draw one triangle with an area of 12 cm^2. This may give the children ideas of other triangles they can draw that also have an area of 12 cm^2.

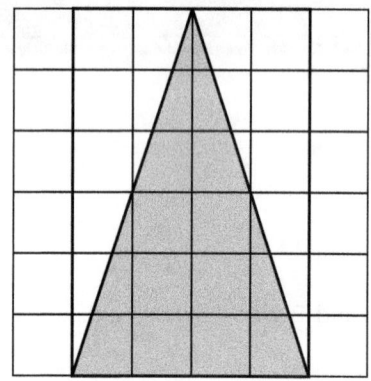

Construct

- You may wish to provide the children with a selection of different envelopes.
- Encourage the children to be as detailed and as accurate as possible with their written instructions, measurements and diagrams.

Page 4

The Puzzler

- Point out that the two square fields need not be the same size.
- Suggest the children draw a diagram, for example:

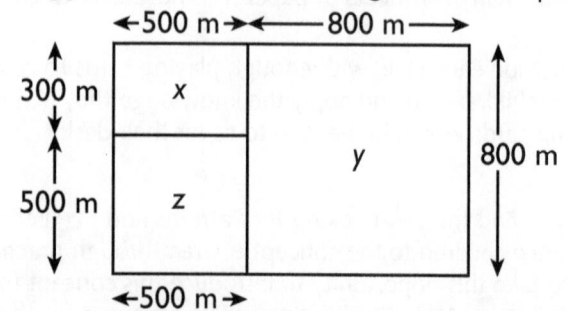

- The farmer kept for himself the rectangular field x with dimensions 300 m × 500 m. Given that the other two fields (y and z) are both squares their dimensions must be as shown above.

Let's Investigate
- Encourage the children to be as accurate as possible when making their comparisons. You may need to discuss with the children how they are going to express their comparisons, for example, using fractions.
- Suggest the children fold the newspaper into eighths or sixteenths, then draw lines and make comparisons.

What's the Problem?
- Ensure children understand that although it is a six-day journey to cross the desert, Roshan takes much longer because she can carry only four litres of water at a time.
- Draw a "map" of the desert divided into six parts to mark the six days' journey. Let children use a cube for Roshan and counters for the litres of water, to help them find a way of crossing the desert.
- Suggest children work in pairs to discuss ideas and share their reasoning.

AfL

- What patterns are there on a calendar?
- What can you tell me about area?
- Talk me through some of the shapes you drew. How are they similar? How are they different?
- How could you improve on your set of instructions?
- What did you use to help you solve this problem / puzzle?
- Tell me what you found out during the course of your investigation.

Issue 25 – Measurement

Answers

Page 1

What's the Problem?
Trevor is 5 years old.

Let's Investigate
Opposite totals are the same.
Other observations may include: the difference between the top left number and bottom right number is always 8, and the difference between the top right number and bottom left number is always 6.

Page 2

Let's Investigate
Results of the investigation will vary.

Let's Investigate
Results of the investigation will vary.

Page 3

Looking for Patterns
Other rectangles with an area of 36 cm² include: 6 × 6, 12 × 3, 18 × 2, 36 × 1

Let's Investigate
Results of the investigation will vary.

Construct
Envelopes and instructions will vary.

Page 4

The Puzzler
The farm was 800 m × 1300 m.

Let's Investigate
Results of the investigation will vary.

What's the Problem?

Journey 1
Roshan sets out from Base Camp and walks for one day into the desert with 4 l of water. She drinks 1 l on the way out and spends the night at Camp 1. The next day she leaves 2 l of water behind at Camp 1 and her fourth litre she drinks on the journey back to Base Camp.

Journey 2
Roshan walks one day into the desert with 4 l of water. She drinks 1 l on the way. She spends the night at Camp 1. The next day she picks up 1 l of the water she had left behind previously, and journeys on into the desert with 4 l. During the day she drinks 1 l. At the end of the second day she is two days journey into the desert and has 3 l of water. She spends the night at Camp 2. The third day she leaves 2 l of water behind at Camp 2 and drinks her fourth litre as she walks back to Camp 1. The fourth day she picks up the second litre of water that she had left at Camp 1 and walks back to Base Camp. There are now 2 l of water left at Camp 2 and none at Camp 1.

Journey 3
Roshan sets out from Base Camp with 4 l of water and spends two days walking to Camp 2. During the two days she drinks 2 l of water and arrives at Camp 2 with 2 l left. On the third day she sets off from Camp 2 with the 2 l of water she had brought with her, plus the 2 l she had left behind on Journey 2, 4 l in total. Ahead of her is four days' journey to the end of the desert.

Inquisitive ant

area
The amount of surface space enclosed within a boundary (perimeter).

Issue 26
Measurement

Prerequisites for learning

- Identify patterns and relationships involving numbers
- Calculate mentally with whole numbers
- Develop and use written methods to add, subtract, multiply and divide whole numbers
- Read, to the nearest division and half-division, scales that are numbered or partially numbered
- Make estimations and approximations
- Use a calendar

Resources

pencil and paper
Resource sheet 2: My notes (optional)
Resource sheet 3: Pupil self assessment booklet (optional)
Resource sheet 8: 1 cm squared paper
ruler
selection of different newspapers
tape measure
metre rule
trundle wheel
computer

Teaching support

Page 1

Looking for Patterns
- The dimensions of the playground are irrelevant to solving this problem.
- Suggest the children draw a picture, for example:

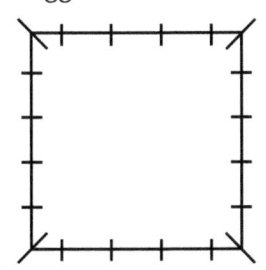

The Language of Maths
- Children provide a definition for each word on their list, in the context of weather.

What's the Problem?
- In 10 minutes the water tap fills the container $\frac{10}{30} = \frac{1}{3}$ full. In 10 minutes the juice tap fills the container $\frac{10}{60} = \frac{1}{6}$ full. Together, in 10 minutes they fill the container $\frac{1}{3} + \frac{1}{6} = \frac{2}{6} + \frac{1}{6} = \frac{3}{6} = \frac{1}{2}$ full. Therefore it takes 2×10 minutes = 20 minutes to fill the container fully.
- What fraction of the liquid in the container is pear juice and what fraction is water?

Page 2

Around the World
- This activity focuses on two aspects of measuring – length (distance) and time.
- Before the children work independently on this activity, you may wish to discuss with the children the five aspects of the data handling cycle:
    Step 1: Plan
    Step 2: Collect data
    Step 3: Process the data
    Step 4: Represent the data
    Step 5: Interpret and discuss the data

173

Issue 26 – Measurement

What's the Problem?
- Suggest the children draw diagrams to assist them in working out the answer, for example:

90 cm

- Ask the children to write a calculation to express the answer, i.e.
  (4 × 40 cm) – 70 cm
  = 160 cm – 70 cm
  = 90 cm

Page 3
Let's Investigate
- This activity, and the next Looking for Patterns activity informally introduce children to the concept of perimeter through practical, discovery-based learning activities. You may wish to take this opportunity to introduce this concept to the children more formally once they have completed these two activities. Note that the Inquisitive ant word for this issue is 'perimeter'.
- Children can use either a tape measure or a trundle wheel to take measurements.
- Children use the dimensions to calculate the area of each space.

Looking for Patterns
- This activity, and the previous Let's Investigate activity informally introduce children to the concept of perimeter through practical, discovery-based learning activities. You may wish to take this opportunity to introduce this concept to the children more formally once they have completed these two activities. Note that the Inquisitive ant word for this issue is 'perimeter'.
- If the children have previously completed Issue 25 or already have some understanding of the concept of area, then ask the children questions similar to the following:
  – What is the area of each of your rectangles? (6 cm × 6 cm: area = 36 cm^2; 7 cm × 5 cm: area = 35 cm^2; 9 cm × 3 cm: area = 27 cm^2; 10 cm × 2 cm: area = 20 cm^2; 11 cm × 1 cm: area = 11 cm^2)
  – Which rectangle has the smallest area? (11 cm × 1 cm)
  – Which rectangle has the largest area? (6 cm × 6 cm)

Focus on Science
- This is an extremely open-ended activity that requires children to think about the organisation of an Earth calendar and apply similar rules and patterns to creating a calendar for an alien planet. However, just as the Earth calendar has different patterns, and exceptions (i.e. the number of days in each month varies in an Earth calendar), so too will the children's Xavion calendar. The important aspect of this activity is the explanations and reasoning the children give for the patterns (and exceptions) for their calendar.
- You may wish to discuss with the children that one of the reasons for the irregularity in the Earth's calendar is due to the fact that there are actually 356.24 days in a year and that this is why every four years we have a leap year to catch up.
- It is recommended that children work in pairs for this activity so that they can share ideas and discuss their reasoning.
- You may wish to ask the children to create their calendar using ICT.

Page 4
The Language of Maths
- Discuss with the children how they should consider instruments that measure length / distance, mass, capacity, time and temperature.

- Children should also think of other instances where scales are used, often where intervals represent non-numerical values, for example: mobile phones (battery life and signal strength); electrical appliances (washing machines and dishwasher cycles, food processor speeds and fridge temperatures); sound producing equipment with volume controls (MP3 players, CD players and television; and cars (fuel gauge, speedometer and tyre pressure gauge).

Around the World
- Ensure children understand how to read and interpret a concentric circle map.

AfL

- What information was important for solving this problem? What information did you *not* need to use?
- Tell me some words that are used to describe the weather.
- Are either of these statement true? When might they be false?
- What can you tell me about perimeter?
- What are the same / different about your shapes?
- Tell me about your Xavion calendar. How does it work? What is regular about it? What is irregular?
- Tell me some of the stranger instruments that you found that use a scale. What do these instruments do? How do the scales work? What do they tell us?
- Look at this map. Explain to me how it works. Tell me the distance between two different towns on the map.

Answers

Page 1

Looking for Patterns
There are six posts on each side of the fence.

The Language of Maths
Lists will vary. However, they should include words such as: higher, lower, greater than, less than, increase, decrease, rise, fall, minus, plus.

What's the Problem?
It takes 20 minutes to fill the container using both taps.

Page 2

Around the World
Results of the investigation will vary.

What's the Problem?
The fewest number of jumps is five jumps: four jumps of 40 cm forward and one jump of 70 cm back.

Page 3

Let's Investigate
Results of the investigations will vary.

Looking for Patterns
Other rectangles with perimeters of 24 cm include:
6 cm × 6 cm
7 cm × 5 cm
9 cm × 3 cm
10 cm × 2 cm
11 cm × 1 cm

Focus on Science
Calendars will vary.

Page 4

The Language of Maths
Lists will vary.

Around the World
Statements will vary.

Inquisitive ant

perimeter
The distance all the way round something – the edge or boundary.

Issue 27
2-D shapes

Prerequisites for learning

- Identify patterns and relationships involving shapes
- Visualise common 2-D shapes
- Identify shapes from pictures of them in different positions and orientations
- Sort, make and describe shapes, referring to their properties
- Identify right angles in 2-D shapes
- Read and record the vocabulary of position
- Solve logic puzzles

Resources

pencil and paper
Resource sheet 2: My notes (optional)
Resource sheet 3: Pupil self assessment booklet (optional)
ruler
scissors
geometric shapes (circle, triangle, square and rectangle) of different sizes
two sheets of A4 paper
matchsticks
art straws and modelling clay (or similar)
geoboards and elastic bands
computer with Internet access

Teaching support

Page 1

The Language of Maths

- Remind the children of the word "regular" as it relates to 2-D shapes. The use of this term and precise positional language is important to describing the design or picture successfully.

The Puzzler

- Children should have had experience of other logic puzzles before attempting this puzzle.
- Children create their own 2-D shape logic puzzle for a friend to solve.

Page 2

The Puzzler

- Tell the children that for the second puzzle, there are two different ways to remove four matchsticks to leave six triangles. Can they find them both?
- Children make some matchstick puzzles of their own.

Looking for Patterns

- There are nine small squares (1 × 1), four medium squares (2 × 2) and one large square (3 × 3), making a total of 14 squares altogether.
- Start by asking the children to draw a 2 × 2 square and find out how many squares there are.
- How many squares are there altogether in a 4 × 4 grid?
  (30: 16 (1 × 1 squares), 9 (2 × 2 squares), 4 (3 × 3 squares), 1 (4 × 4 squares).)

Looking for Patterns

- Ensure children are familiar with the term "equilateral triangle".
- What if you used nine art straws and modelling clay (or similar) to make a three-dimensional shape? (Seven triangles makes a double tetrahedron.)

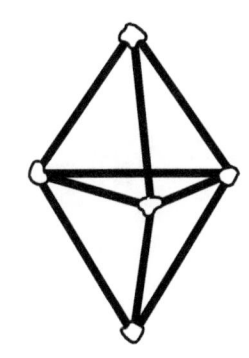

Page 3
The Puzzler
- Point out to the children that the lines can extend beyond the dots.

Let's Investigate
- Suggest the children sort their shapes using the following criteria:
  - triangles, quadrilaterals, pentagons, hexagons
  - shapes with no right angles, shapes with one right angle, shapes with more than one right angle
  - shapes with parallel sides, shapes without parallel sides.
- Once children have made and sorted their shapes, suggest they exchange their set of shapes with the shapes of another child. Children sort each other's shapes. They can then discuss and compare the criteria used to sort each set of shapes.
- Pairs of children combine and sort their shapes. How many different ways can they sort the shapes?

Page 4
Let's Investigate
- Ensure children are familiar with the terms "regular", "irregular" and "polygons".
- Allow the children to use geoboards and elastic bands to construct their polygons.
- When drawing a conclusion about the maximum number of right angles that different polygons can have, you may wish the children to work in pairs so that they can share ideas and discuss their reasoning.

The Arts Roundup
- You may wish to suggest to the children that they print a copy of the pictures and draw around the squares.
- Often the squares in the pictures may not be exactly square, but allow children to count them if they can justify their inclusion.

AfL

- What words were important in writing your description?
- How successful was your friend in recreating your diagram? How could you improve your list of instructions?
- What did you visualise as you were working out this puzzle?
- How did you sort your shapes? How else could you sort them? Any other ways?
- What polygons could you not make? Why?
- Tell me about the conclusions you drew about the maximum number of right angles that different polygons can have.
- What can you tell me about the mathematics behind the work of Piet Mondrian?

Answers

Page 1

The Language of Maths
Diagram and instructions will vary.

The Puzzler

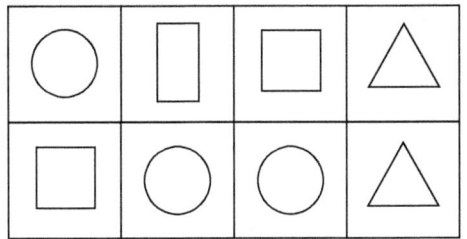

Page 2

The Puzzler

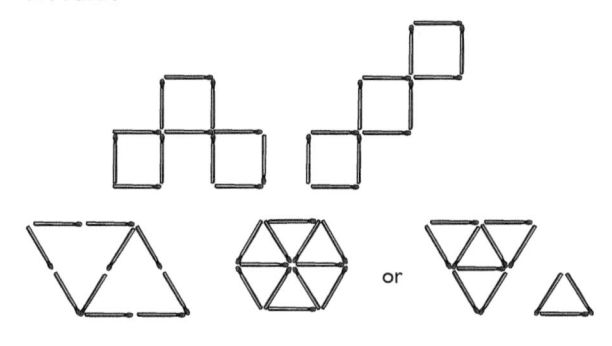

Looking for Patterns
There are 14 squares altogether.

Looking for Patterns
Five equilateral triangles:

Page 3

The Puzzler
Start

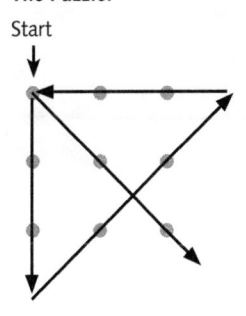

Let's Investigate
Results of the investigation will vary.

Page 4

Let's Investigate
As long as all the angles in the polygon are less than 180°, the maximum number of right angles that it can have is always two less than the number of its sides.

The Arts Roundup
There are 2 squares and 15 rectangles.
Results of the investigation will vary.

Inquisitive ant

regular polygon

A *polygon* is a 2-D shape with three or more straight sides and angles.
A *regular polygon* has sides of equal length and all its angles are equal.

Issue 28

3-D shapes

Prerequisites for learning

- Visualise common 3-D shapes
- Identify shapes from pictures of them in different positions and orientations
- Sort, make and describe shapes, referring to their properties

Resources

pencil and paper
Resource sheet 2: My notes (optional)
Resource sheet 3: Pupil self assessment booklet (optional)
Resource sheet 8: 1 cm squared paper
Resource sheet 9: 2 cm squared paper
Resource sheet 12: Triangular dot paper
Resource sheet 13: Isometric paper (optional)
ruler
scissors
sticky tape
glue
junk construction material
blank dice (or white cubes) and red and blue washable felt-tip pens (optional)
interlocking cubes

Teaching support

Page 1

Construct

- Do the children realise that the number of blocks in the bottom row is the same as the number of rows?

Construct

- The purpose of this activity is for the children to be as creative as possible in both finding and creating examples of different polyhedrons.
- Children may need assistance seeing how existing shapes can be adapted, e.g. a corner of a cardboard box cut off and a triangle of cardboard stuck on the bottom to make a tetrahedron.

Page 2

At Home

- Once the children have completed the activity, ensure that there is an opportunity in class for pairs or groups of children to discuss their results.

What's the Problem?

- Ensure children understand the meaning of the word "cubic".
- Remind the children that a cube has eight corners or vertices.
- How many pieces have rind on only one side?

Construct

- Provide the children with some blank dice (or white cubes) and red and blue washable felt-tip pens.
- How many different ways can you colour the faces of a cube if two faces must be red, two faces blue and two faces green? (There are five ways, six if you count a mirror image.)

Issue 28 – 3-D shapes

Page 3

Construct

- Children investigate how many different cuboid shapes can be made with 18 interlocking cubes.
- Children investigate how many different cuboid shapes can be made with 36 interlocking cubes.
- Ask the children to draw their shapes on isometric paper.

Construct

- Triangular dot paper (Resource sheet 12) is actually triangular dot isometric paper. Once the children have drawn their model using triangular dot paper, ask them to reproduce it on Resource sheet 13: Isometric paper.
- If enough children have completed this activity, collect the drawings (but not the models) from the children and redistribute them. Can the children use the drawings to construct the 3-D models from interlocking cubes?

Page 4

Construct

- Ask questions such as: "How many cubes high is the model?", "Is every part of the model the same height?", "How many cubes are at the base of the model?"
- Allow the children to work in pairs to construct the model.

Construct

- If necessary, allow the children to use interlocking cubes to construct the model and use this to gain the three different perspectives needed to draw the three different views.

Construct

- Ensure children have successfully completed the first two Construct activities on this page before attempting this activity.

AfL

- How did you work out the solution to this problem? Show me your working out.
- Were you able to find a shape with one / two / ... eight faces, or did you have to make one? How did you make it?
- What are the most common shapes used for packaging in supermarkets? Why do you think this is?
- Were you able to visualise the answer to this problem / puzzle or did you use some type of apparatus? How did that help?

Issue 28 — 3-D shapes

Answers

Page 1

Construct
There would be 15 blocks in the bottom row.
There would be 15 rows in the pyramid.

Construct
Constructions will vary.

Page 2

At Home
Cuboids are used for solid goods, because they stack tightly together.
Cylindrical shapes are used for liquids, because they hold the most liquid with the least packaging, but still stack closely.

What's the Problem?
Eight of the smaller cubes have rind on three sides.

Construct
There are two different ways possible:
R = Red   B = Blue
Other solutions are possible, however, they are all variations of those shown here.

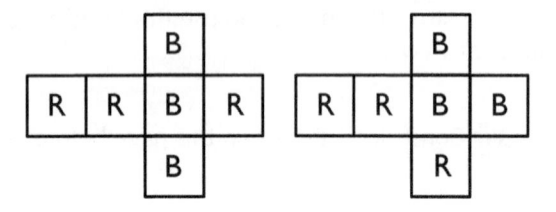

		B	
R	R	B	R
		B	

		B		
R	R	B	B	
		R		

Page 3

Construct
24 cubes can be arranged into six different cuboids:
1 × 1 × 24, 1 × 2 × 12, 1 × 3 × 8, 1 × 4 × 6, 2 × 2 × 6 and 2 × 3 × 4.

Construct
Models and drawings will vary.

Page 4

Construct

Construct
Front view   Top view   Side view

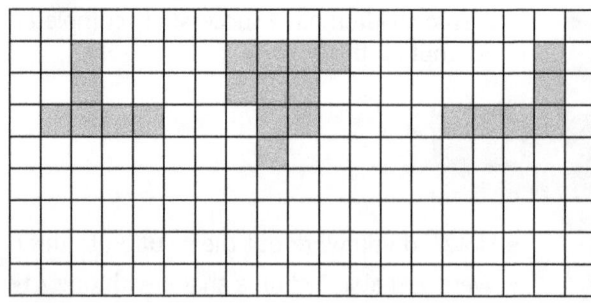

Construct
Models and drawings will vary.

Inquisitive ant

net
A 2-D shape that can be folded up to make a 3-D shape.

182

Issue 29
Symmetry

Prerequisites for learning

- Identify reflective symmetry in patterns and 2-D shapes
- Draw lines of symmetry in shapes
- Draw and complete shapes with reflective symmetry
- Draw the reflection of a shape in a mirror line along one side
- Read and record the vocabulary of position, direction and movement

Resources

pencil and paper
Resource sheet 2: My notes (optional)
Resource sheet 3: Pupil self assessment booklet (optional)
Resource sheet 4: Symmetrical circles
Resource sheet 8: 1 cm squared paper
Resource sheet 9: 2 cm squared paper
Resource sheet 10: 0·5 cm squared paper
Resource sheet 12: Triangular dot paper
Resource sheet 13: Isometric paper (optional)
ruler
interlocking cubes
coloured pencils
camera
range of different measuring equipment
scissors
glue
computer

Teaching support

Page 1

Construct

- Ask the children to draw one or more of their models on isometric paper, using colour to show symmetry.

Let's Investigate

- This activity is best done in pairs or small groups.
- Allow the children to make a display of their photos. Alongside each photo they should include a brief comment on how each one shows reflective symmetry.
- Children find examples in the classroom and around the school that show rotational symmetry.

Page 2

Famous Mathematicians

- Remind the children to be as accurate as possible in their construction of their Koch Snowflake.
- The main focus of this activity is the use of colour to show reflective symmetry.

Let's Investigate

- Children will need to work in pairs for this activity. It is recommended that children complete this and the Let's Investigate activity on page 3 together.
- Children will also need to use a range of different measuring equipment. However, do not suggest which equipment the children should use. Let them discover this for themselves during the course of the investigation.
- Discuss with the children how they are going to record their results and how they are going to justify their conclusion as to whether or not their face is symmetrical.

Page 3

Let's Investigate

- Children will need to work in pairs for this activity. It is recommended that children complete this and the Let's Investigate activity on page 2 together.
- Children will also need to use a range of different measuring equipment. However, do not suggest which equipment the children should use. Let them discover this for themselves during the course of the investigation.
- Discuss with the children how they are going to record their results and how they are going to justify their conclusion as to whether or not their body is symmetrical.

The Language of Maths

- Encourage children to check that they have drawn the mirror image route correctly before they move on to the next two stages of the activity.

Page 4

Construct

- Children can use 0·5 cm, 1 cm or 2 cm squared paper for this activity.
- Ensure that the children are accurate in both their cutting out of the circles and quarters, as well as in placing their quarters onto the squared paper.
- You may need to provide children with an example to get them started with this investigation.

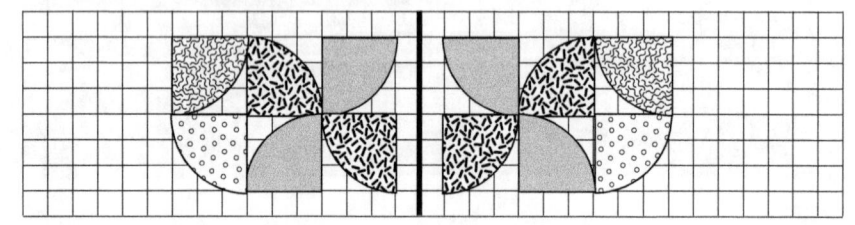

- Suggest the children create patterns where:
– the edges of the shape are parallel or perpendicular, but not touching the mirror line, for example:

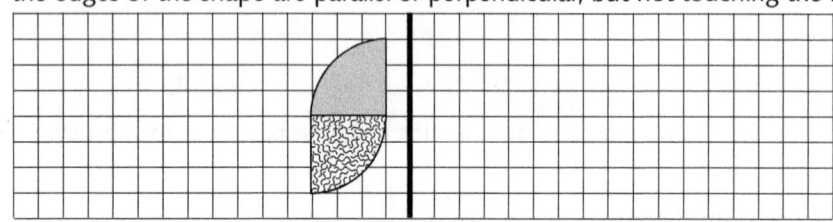

– a corner of the shape is touching the mirror line, for example:

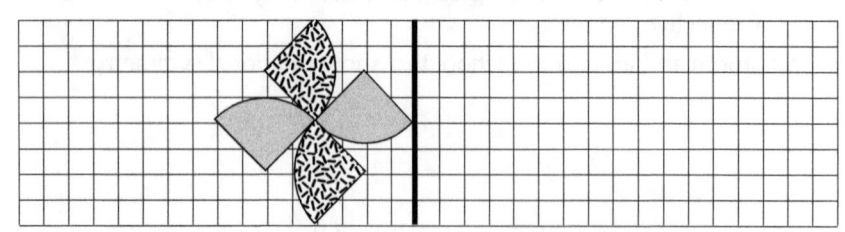

Looking for Patterns

- Ensure children realise that there are two parts to this activity. Firstly, to identify and continue the pattern in Grids 4 and 5; secondly, to find pairs of grids that, when placed side by side, show reflective symmetry.
- Children make their own set of five 5 × 5 grids (or seven 7 × 7 grids). In the first three grids they show a repeating pattern by translating the coloured squares horizontally. They then ask a friend to complete the remaining grids and identify the pairs of grids that, when placed side by side, show reflective symmetry along a vertical line of symmetry.
- Children make their own set of grids and translate the coloured squares vertically. They then ask a friend to complete the remaining grids and identify the pairs of grids that, when placed one on top of another, show reflective symmetry along a horizontal line of symmetry.

AfL

- How does your model / diagram show symmetry?
- Talk me through your collection of photographs and how they show reflective symmetry.
- How symmetrical are you?
- The moving pattern of shaded squares is called a translating pattern. Describe for me how the shaded squares in *this* grid have been moved to create the shaded squares in *this* grid.

Issue 29 — Symmetry

Answers

Page 1

Construct
Models and statements will vary.

Let's Investigate
Results of the investigation will vary.

Page 2

Famous Mathematicians
Koch Snowflakes and reports on symmetry will vary.

Let's Investigate
Results of the investigation will vary.

Page 3

Let's Investigate
Results of the investigation will vary.

The Language of Maths
Where do fish keep their money? In a riverbank.

Page 4

Construct
Patterns will vary.

Looking for Patterns

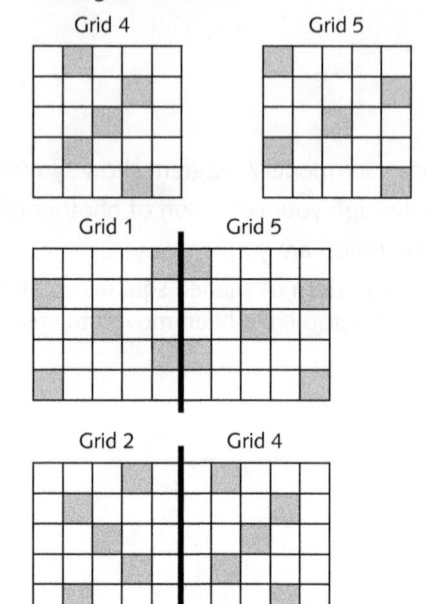

Grid 4   Grid 5

Grid 1 | Grid 5

Grid 2 | Grid 4

Inquisitive ant

rotational symmetry
Rotation involves the circular movement of a shape about a given point or line (axis) without changing the shape. A shape has *rotational symmetry* if it looks the same after a certain amount of rotation.

Issue 30
Position and direction

Prerequisites for learning

- Use mathematical vocabulary to describe position, direction and movement, including movement in a straight line and distinguishing between rotation as a turn and in terms of right angles for quarter, half and three-quarter turns (clockwise and anti-clockwise)
- Know that one right angle is 90°
- Describe movement about a grid
- Solve logic puzzles

Resources

pencil and paper
Resource sheet 2: My notes (optional)
Resource sheet 3: Pupil self assessment booklet (optional)
Resource sheet 5: Draughts board
Resource sheet 8: 1 cm squared paper
Resource sheet 11: Squared dot paper
ruler
red, blue, green, yellow, black and purple coloured pencils
"small treasure"
red, green, blue and yellow counters (optional)
four draughts pieces and four identical chess pieces, e.g. pawns (optional)

Teaching support

Page 1

The Language of Maths

- Encourage the children to be as detailed as possible in both the drawing of their map and the writing of their clues, and to use precise mathematical language for position and direction.
- Children hide the treasure somewhere within the school.

The Puzzler

- Children should have had experience of other logic puzzles before attempting this puzzle.

Page 2

Technology Today

- Ensure children are familiar with a floor robot and the term "Turn left 90°".

Technology Today

- Ask the children to write another path for the robot. Children swap their program and draw the route on squared dot paper.

What's the Problem?

- Suggest the children draw a diagram to help them work out the answers, for example:

Louisa's Home   Helen's Home   Jason's Home

or

Helen's Home   Louisa's Home   Jason's Home

- Tell the children that there are two possible answers to this problem. Can they find them both?

Page 3

At Home

- You may need to remind children about the four compass directions and how to estimate direction using the sun and locally known features.
- Once the children have completed the activity, ensure that there is an opportunity in class for pairs or groups of children to discuss their results.

Looking for Patterns

- When using the Resource sheet to record their results, suggest the children write "D" for a draughts piece and "C" for a chess piece or draw a circle for the draughts piece and a vertical line for the chess piece.
- Provide the children with four draughts pieces and four identical chess pieces, e.g. pawns.

Page 4

The Puzzler

- The friends are seated, going clockwise, Anna, Dennis, Bashir and Cathy. Cathy is on Anna's right so she is using yellow. Dennis isn't using blue, so he must be using green, which leaves Bashir using blue.
- Provide the children with red, green, blue and yellow counters.

The Language of Maths

- Suggest children draw their route on the grid as they go along.
- Children create a set of directions for making certain words using the letters provided on the grid, e.g. number, division, total, estimate, length, circle, mathematics, ...

AfL

- Talk me around your map. Let me see if I can use your set of clues to see where you hid the map.
- Tell me how you worked out the solution to this puzzle.
- Show me your program for the robot. Let's follow the program together on the path that shows the robot's route.
- Were you able to visualise where the chess and draught pieces belonged on the board or did you use actual chess and draught pieces?
- Beginning at "start", tell me the directions you would need to follow to spell the word "clever".

Answers

Page 1

The Language of Maths
Maps and clues will vary.

The Puzzler

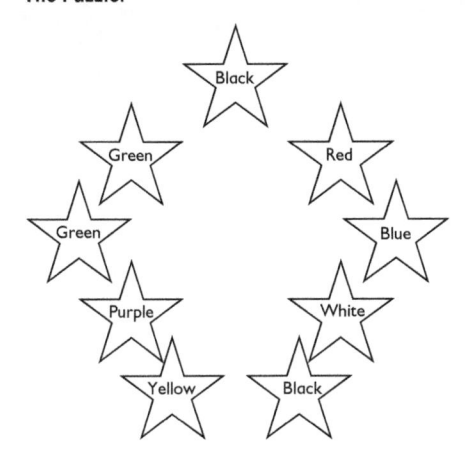

Page 2

Technology Today

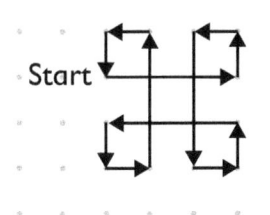

Technology Today

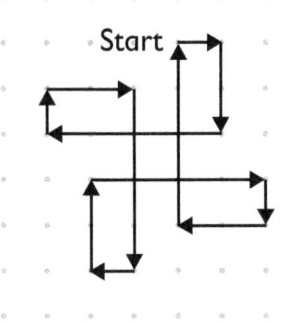

Children's own routes will vary.

What's the Problem?
There are either four or 12 houses between Jason's home and Helen's home.

Page 3

At Home
Generally, churches will face east and mosques towards Mecca.

Looking for Patterns
D = Draughts piece
C = Chess piece

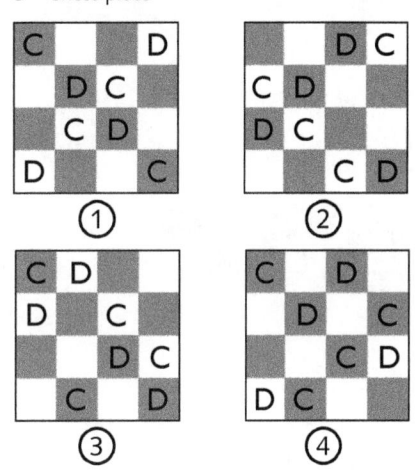

Page 4

The Puzzler
Anna – red, Dennis – green, Bashir – blue, Cathy – yellow

The Language of Maths
What makes an octopus laugh? Ten tickles.

Inquisitive ant

° degrees
A unit of measurement of angles.

Issue 31

Movement and angle

Prerequisites for learning

- Use mathematical vocabulary to describe position, direction and movement, including movement in a straight line and distinguishing between rotation as a turn and in terms of right angles for quarter, half and three-quarter turns (clockwise and anti-clockwise)
- Know that one right angle is 90°
- Describe movement about a grid

Resources

pencil and paper
Resource sheet 2: My notes (optional)
Resource sheet 3: Pupil self assessment booklet (optional)
Resource sheet 8: 1 cm squared paper
Resource sheet 9: 2 cm squared paper
Resource sheet 12: Triangular dot paper
ruler

Teaching support

Page 1

At Home

- Once the children have completed the activity, ensure that there is an opportunity in class for pairs or groups of children to discuss their results.

Focus on Science

- Ensure children are familiar with the language used in the activity, in particular the terms "quarter turn", "half turn", "right angle", "clockwise" and "anticlockwise".

Page 2

The Puzzler

- Ensure the children realise that each set of instructions is written from the perspective of someone reading the issue, and not from the perspective of someone walking through the maze.
- Ask the children to write a set of instructions from the perspective of someone continually walking through the maze and turning according to the directions given.

The Language of Maths

- Discuss with the children the different mathematical terms they could use to describe the route through the maze, for example, "clockwise" / "anticlockwise", "left" / "right", "90 degrees" / "through one right angle", …

Page 3

Looking for Patterns

- Ensure children understand the term "clockwise 90°". Discuss how this movement can also be referred to as "moving through one right angle to the right" or as "a quarter turn to the right".

Looking for Patterns

- Encourage the children to reflect the lines at an angle other than 90°, for example:

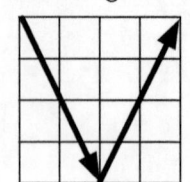

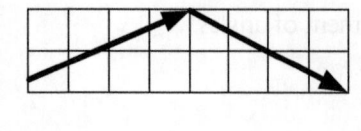

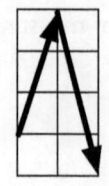

- Children investigate what patterns they can make using triangular dot paper.

190

Page 4

Around the World
- Tell the children that there are a total of ten different routes possible.
- Children can draw their routes using either 1 cm or 2 cm squared paper.

The Language of Maths
- This activity introduces children to the concept of mathematical translations (both one and two translations).
- If children have previously completed the Looking for Patterns activity on page 4 of Issue 29 – Symmetry – you may wish to discuss with them how this activity also involved translations.
- Children draw a simple shape on squared paper and translate the shape using one or two translations. Encourage them to label each of the translated shapes, and write how they have been translated using the language "to the left", "to the right", "up" and / or "down".

AfL

- Give me some examples of different things that open and close. What can you tell me about the different angles that are created?
- How else can you describe a quarter turn to the right? What about two right angles anticlockwise?
- How good is your set of instructions for getting out of the maze? Can you improve on them? How?
- Where would the star be in the next shape in the sequence? How do you know that?
- Describe your pattern to me.
- Which route do you think would be the fastest for Ben to take to get from his home to school?
- Describe for me how Shape A has been translated to Shape D.
- How has Shape D been translated from Shape C?

Issue 31 – Movement and angle

Answers

Page 1

At Home
Results of the investigation will vary.

Focus on Science
1. D   2. C   3. D   4. C   5. B   6. C

Page 2

The Puzzler

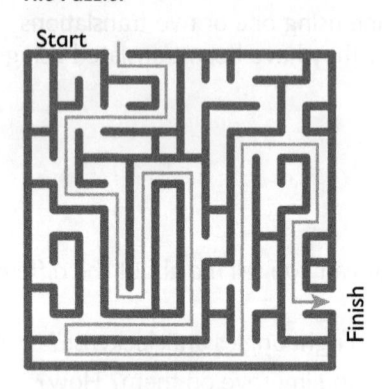

The Language of Maths
Routes and instructions will vary.

Page 3

Looking for Patterns

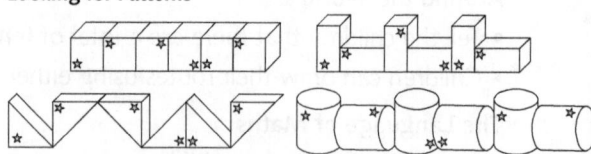

Looking for Patterns
Children's patterns will vary 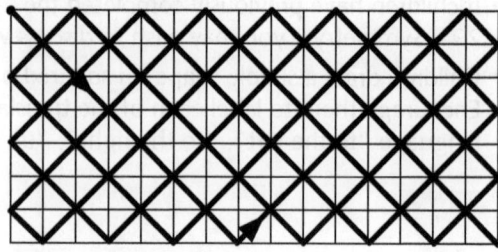

Page 4

Around the World

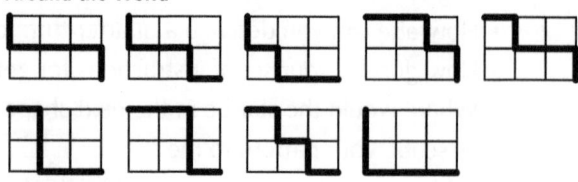

The Language of Maths

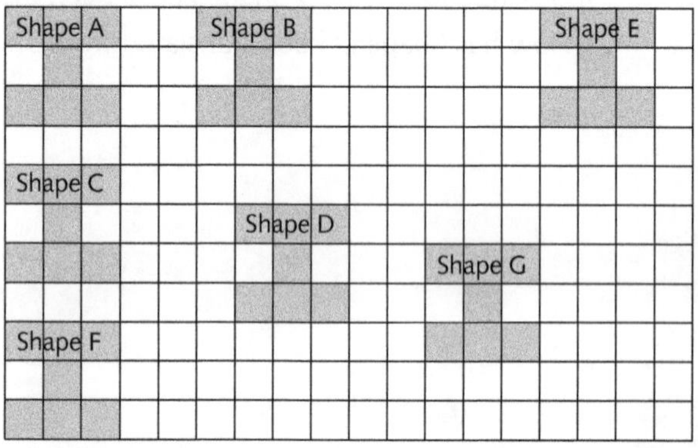

Inquisitive ant

translation
The movement, or sliding, of a shape in a certain direction – up, down, left or right – without changing its shape in any way.

192

Issue 32
Geometry

Prerequisites for learning

- Identify patterns and relationships involving shapes
- Visualise common 2-D shapes and 3-D shapes
- Identify shapes from pictures of them in different positions and orientations
- Sort, make and describe shapes, referring to their properties
- Use mathematical vocabulary to describe position, direction and movement
- Use the four compass directions to describe movement about a grid

Resources

pencil and paper
Resource sheet 2: My notes (optional)
Resource sheet 3: Pupil self assessment booklet (optional)
Resource sheet 6: Half a shape
Resource sheet 8: 1 cm squared paper
Resource sheet 9: 2 cm squared paper
Resource sheet 13: Isometric paper
ruler
scissors
interlocking cubes
set of 1–20 number cards (optional)

Teaching support

Page 1

Construct

- These are the five different four-cube shapes that are one cube deep.

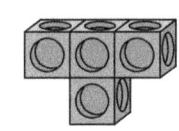

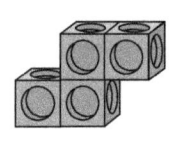

Looking for Patterns

- When recording which shapes, when duplicated four times, will fit together to make a 4 × 4 square, suggest the children use colour. For example:

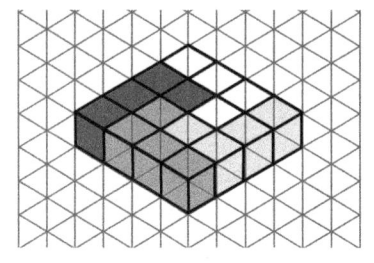

Page 2

Let's Investigate

- Discuss with the children how shapes such as the following are considered to be the same.

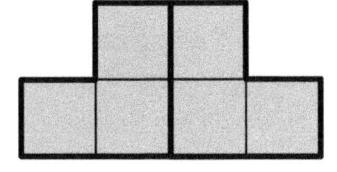

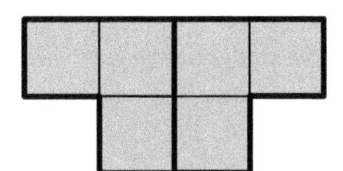

- Ensure the children realise that when forming their shapes that adjoining squares must all align, for example.

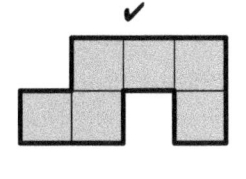

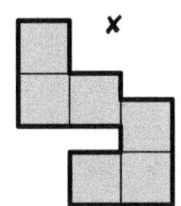

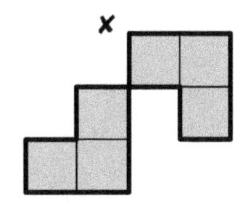

193

Issue 32 – Geometry

- Do not provide the children with the Resource sheet. Instead, encourage them to visualise how two of each of the figures makes a whole shape. Children still record their results on squared paper.
- What if the figures on the Resource sheet were only one-quarter of the shape?

Let's Investigate

- Once children have completed the activity, arrange children into pairs and ask them to compare their diagrams. Which diagrams are the same? How many different diagrams are they able to make between them?

Looking for Patterns

- There are nine small triangles (△), three medium triangles (△△) and one large triangle, making a total of 13 triangles altogether.

Page 3

The Puzzler

- Give children an enlarged copy of the page from the Issue and let them cut out the shapes and fit them together.

The Puzzler

- Ask the children to make a similar puzzle of their own for a friend to solve. Can they construct one that uses more than half the squares on the grid? What about constructing a 7 × 7 square puzzle?

Page 4

The Puzzler

- Suggest the children draw and label a circle.
- Some children may find it useful to use a set of number cards to "act out" the problem.

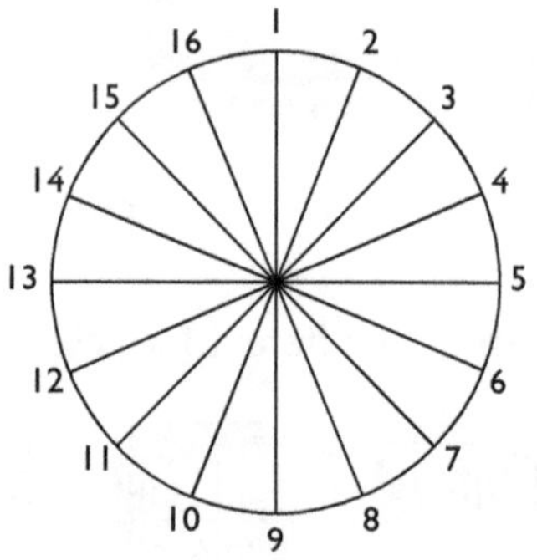

Looking for Patterns

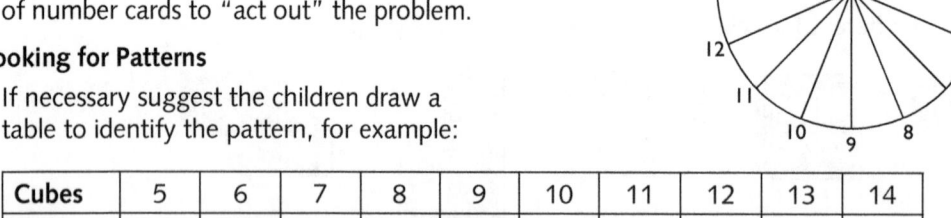

- If necessary suggest the children draw a table to identify the pattern, for example:

Cubes	5	6	7	8	9	10	11	12	13	14
Cars	1	2	3	4	5	6	7	8	9	10

AfL

- This shape you have drawn on the isometric paper, can you make it for me using these cubes?
- Tell me about the different ways you can draw four / five straight lines.
- What can you tell me about the number of intersecting points there are for diagrams drawn with four / five straight lines?
- What system did you use to make sure that you counted all the triangles?
- Which pieces of the jigsaw did you know straightaway were / were not a part of the hexagon?
- What patterns did you recognise that made it easy to make a prediction as to the number of cubes you would need?

Answers

Page 1

Construct
These are the five different four-cube shapes.

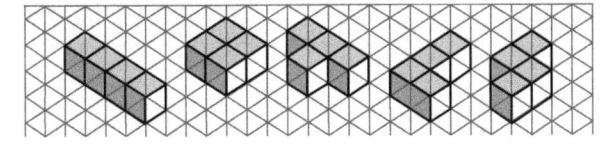

Looking for Patterns
The first four shapes above (i.e. shapes I, O, T and L) will each fit together to make a 4 × 4 square when replicated four times. The final shape above (i.e. shape Z) will not.

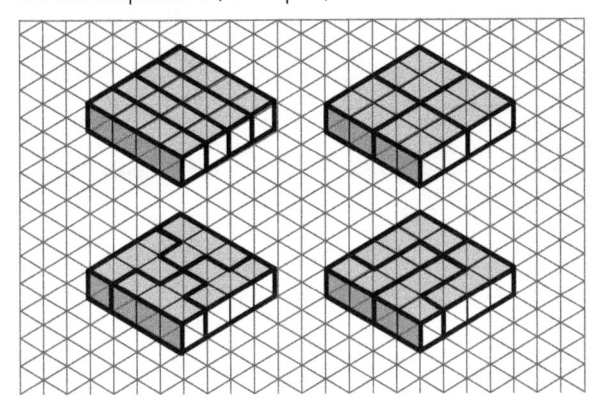

Page 2

Let's Investigate

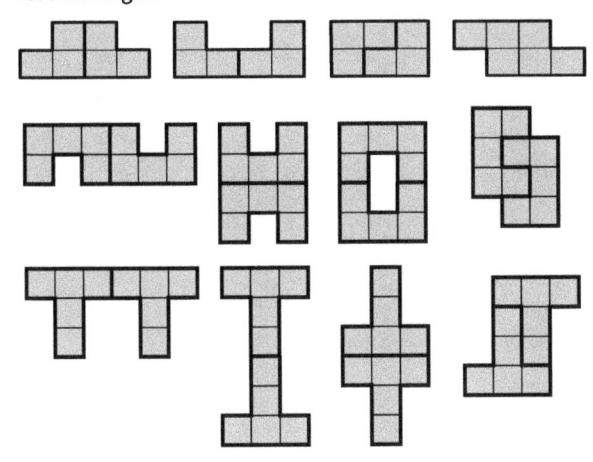

Other shapes are possible.

Let's Investigate

Looking for Patterns
There are 13 triangles altogether.

Page 3

The Puzzler

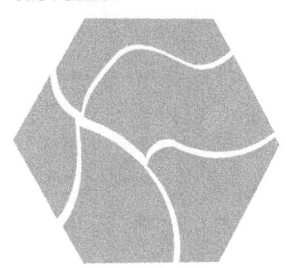

The Puzzler
2N

Page 4

The Puzzler
16 friends are sitting round the table. Children's explanations will vary.

Looking for Patterns
8 cars will go into a garage made of 12 cubes.
18 cubes are needed to make a garage that will fit 14 cars.

Inquisitive ant

intersection
The place or point where two or more things cross each other.

Issue 33
Geometry

Prerequisites for learning

- Identify patterns and relationships involving shapes
- Visualise common 2-D shapes and 3-D shapes
- Identify shapes from pictures of them in different positions and orientations
- Sort, make and describe shapes, referring to their properties
- Use mathematical vocabulary to describe position, direction and movement
- Use the four compass directions
- Solve logic puzzles

Resources

pencil and paper
Resource sheet 2: My notes (optional)
Resource sheet 3: Pupil self assessment booklet (optional)
Resource sheet 7: Tangram
Resource sheet 8: 1 cm squared paper
Resource sheet 9: 2 cm squared paper
ruler
scissors
large square sheet of coloured paper
atlas (optional)

Teaching support

Page 1

The Puzzler

- This is a simplified Sudoku puzzle arranged in a 6 × 6 grid rather than a 9 × 9 grid, and uses shapes rather than digits. Ensure the children know the rules of the puzzle, and if they are familiar with how to solve a standard Sudoku puzzle, make sure that they understand how this puzzle differs.

Construct

- Tell children how many different ways each shape can be made, i.e. 2 × 3 = 2 ways; 3 × 3 = 2 ways; 3 × 4 = 4 ways.
- How many different ways can you make a 3 × 5 rectangle?

Page 2

Construct

- If appropriate, allow the children to work in pairs or groups to make their own origami box.
- Children can investigate other origami shapes.

Page 3

Looking for Patterns

- To work out what shape is on the opposite face to ▣, first ask the children to work out the four shapes on the faces next to ▣ (i.e. ▲, ✘, ◉, ✦). So, ★ is opposite ▣.
- To work out what shape is on the opposite face to ▲, first ask the children to work out the four shapes on the faces next to ▲ (i.e. ▣, ✘, ★, ✦). So, ◉ is opposite ▲.
- To work out what shape is on the opposite face to ✘, first ask the children to work out the four shapes on the faces next to ✘ (i.e. ▣, ◉, ★, ▲). So, ✦ is opposite ✘.

Around the World

- Encourage the children to use the eight, rather than four, compass directions.
- Remind the children that the direction between most places will not follow exactly the four cardinal directions (N, S, E, W) or the four ordinal directions (NE, NW, SE, SW). Children will need to make a judgement as to which direction is the closest.

Around the World
- Some children may need to use an atlas to accurately locate their chosen town or city on the map.

Page 4
In the Past
- Ensure children are able to identify each of the shapes, in particular the parallelogram.
- Children use the tangram pieces to create other shapes and pictures.

The Puzzler
- Draw two identical Greek crosses and cut them out. Can you bisect each of them in the same way so that the four pieces form a large square?

AfL

- How did you find the solution to this puzzle? What strategies did you use?
- Why are these two shapes considered to be the same?
- What mental picture did you have in your mind as you were doing this activity?
- Tell me a place on the map that is north-west of London.
- What things did you think about as you were trying to construct the different shapes using the tangram?

Issue 33 – Geometry

Answers

Page 1

The Puzzler

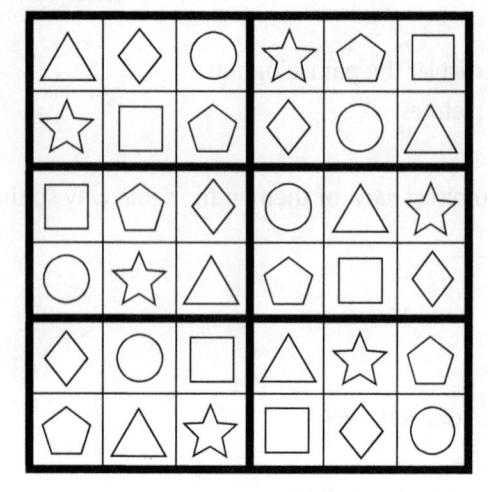

Construct
3 × 2 = 2 ways

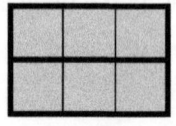

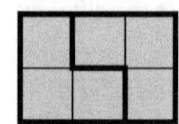

3 × 3 = 2 ways

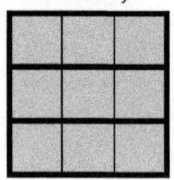

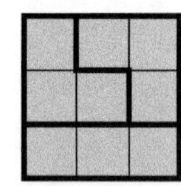

3 × 4 = 5 ways

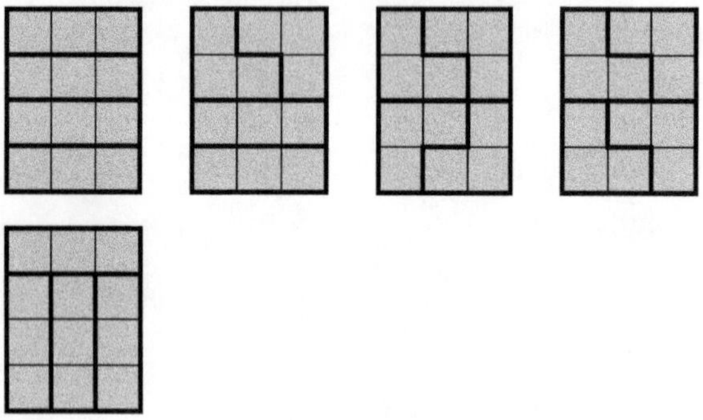

Page 2

Construct
No answer required.

Page 3

Looking for Patterns

The shape on the opposite face to ■ is ★.

The shape on the opposite face to ▲ is ◉.

The shape on the opposite face to ✖ is ✣.

Around the World
Statements will vary.

Around the World
Statements will vary.

Page 4

In the Past

Triangle

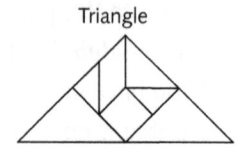

Rectangle

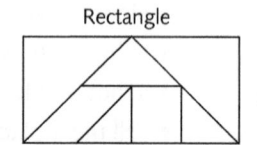

Parallelogram

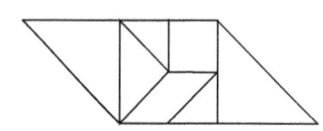

Hexagon

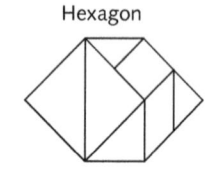

The Puzzler
Wherever the second perpendicular line is drawn dividing the cross into four parts, the parts can be reassembled into a square. If the second line divides the cross into five pieces, they will not form a square.

Inquisitive ant

perpendicular
A line that is *perpendicular* to another line meets it at a right angle.

Issue 34
Statistics

Prerequisites for learning

- Answer a question by collecting, organising and interpreting data
- Use tally charts, frequency tables, pictograms and bar charts to represent results and illustrate observations
- Use Venn diagrams or Carroll diagrams to sort data and objects using more than one criterion
- Make estimations and approximations

Resources

pencil and paper
Resource sheet 2: My notes (optional)
Resource sheet 3: Pupil self assessment booklet (optional)
art paper and colouring materials
computer with Internet access

Teaching support

Page 1

Famous Mathematicians

- Remind the children where to include those children that belong outside the two intersecting rings, i.e. in question 2, the two children who do not like tennis or football (see Answer).
- Ask the children to draw Venn diagrams to represent different classroom situations. Encourage them to be as creative as possible when choosing their criteria.

Around the World

- This activity enables children to appreciate that data handling has a practical and useful application. Through discussion, also highlight to the children the importance of making the data meaningful and accessible to the reader.

Page 2

Let's Investigate

- Children carry out the two surveys on their family, asking as many family members as they can. How do the results for different families differ? Why do the children think this might be?

Let's Investigate

- While the data collection is an important aspect of this activity, the most important part is in the children deciding on a lunch menu that is representative of the most popular lunch in the class.
- Once the children have written their menu, discuss with them why they consider the menu to be representative of the class.
- Children can create their menu using ICT.

Page 3

The Language of Maths

- This activity highlights the different interpretations that different people can have of the same set of data. What is important, however, is that children are able to give a reasonable explanation as to what the bar chart might represent, and recognise and logically explain the trends that the chart displays. Therefore, once the children have completed the activity, conduct a group discussion providing an opportunity for each child to describe their interpretation of the bar chart.

In the Past

- Ensure children know the relationships of the people involved in this question.
- Using different criteria, can you draw another Carroll diagram that includes the same four people?

Page 4

Let's Investigate

- Discuss with the children how they are going to find out how many extended family members each person in the class has. What is the least disruptive way of collecting the data?
- You may also need to discuss with the children how they are going to work out the number of extended family members there are in the "average" family in the class.

At Home

- Once the children have completed the activity, ensure that there is an opportunity in class for pairs or groups of children to discuss their results.
- Children investigate which rooms are used the most / least during the course of a weekend.

AfL

- Talk me through your Venn diagrams.
- Does your brochure include everything someone would need to find out about the tourist attraction?
- Tell me the results of your investigation.
- What conclusions can you draw?
- Why did you choose to record your results in this way?
- What might this graph be about? Tell me a story about the information in this graph.

Answers

Page 1

Famous Mathematicians
1. a. 8
   b. 4

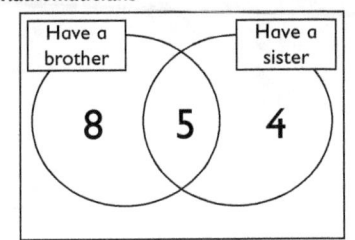

   Venn diagram: Have a brother | Have a sister — 8, 5, 4

2. a. 12
   b. 6

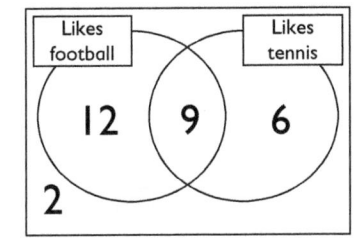

   Venn diagram: Likes football | Likes tennis — 12, 9, 6, 2

Around the World
Brochures will vary.

Page 2

Let's Investigate
Results of the investigation will vary.

Let's Investigate
Results of the investigation will vary.
Lunch menus will vary.

Page 3

The Language of Maths
Labelling and explanations will vary.

In the Past

	Mother	Not mother
Daughter	Queen Elizabeth II	Princess Anne
Not daughter	Queen Elizabeth the Queen Mother	The Prime Minister

Page 4

Let's Investigate
Results of the investigation will vary.

At Home
Results of the investigation will vary.

Inquisitive ant

data

Data is information. It is usually presented as facts or figures and is obtained from experiments or surveys. It is used as a basis for drawing conclusions.

201

Issue 35
Statistics

Prerequisites for learning

- Answer a question by collecting, organising and interpreting data
- Use tally charts, frequency tables, pictograms and bar charts to represent results and illustrate observations
- Use ICT to create a simple bar chart
- Use Venn diagrams or Carroll diagrams to sort data and objects using more than one criterion
- Make estimations and approximations

Resources

pencil and paper
Resource sheet 2: My notes (optional)
Resource sheet 3: Pupil self assessment booklet (optional)
Resource sheet 8: 1 cm squared paper
Resource sheet 9: 2 cm squared paper
ruler
four different coloured pencils, preferably red, orange, green and yellow
dictionary, preferably mathematical
data handling software (optional)
computer (optional)

Teaching support

Page 1

The Language of Maths

- Suggest the children use red (strawberry), orange (orange), green (lime) and yellow (lemon) coloured pencils.
- There are two important aspects to this investigation: firstly, to be able to re-represent the data in the graph in a diagram, and secondly to make statements about this data.

At Home

- Once the children have completed the activity, ensure that there is an opportunity in class for pairs or groups of children to discuss their results.

Page 2

The Language of Maths

- Encourage children to write statements along the lines of "Nine children like boiled eggs and toast".
- Ensure children have completed and understood this activity before starting on the Let's Investigate activity on page 2.

Let's Investigate

- Before children start this activity, discuss with them the different ways they could present their results (including the Venn diagram in The Language of Maths activity above).
- Encourage the children to present their results as both a Venn diagram and a table.

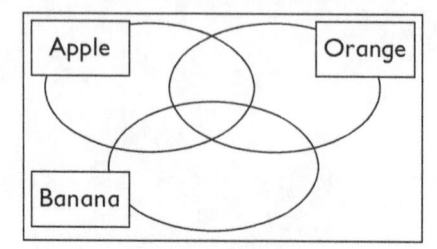

Page 3

Let's Investigate

- Discuss with the children how they are only expected to make an estimate of the total amount of time spent eating. However, they need to ensure that this estimate is as accurate as possible.
- If appropriate, children represent their data using ICT.

The Language of Maths

- Once the children have completed their list, allow them to use a dictionary (mathematical, if available) to check their work, including their spellings.
- You may wish to ask the children to do this activity in pairs.

Focus on Science

- Discuss with the children the different types of material commonly seen in the classroom.
- Also discuss how most objects will not be made solely using one type of material. Chairs, for example, are sometimes made from a combination of plastic and metal (and perhaps even rubber).

Page 4

At Home

- Discuss with the children how, for most children, it will not be possible for them to find out exactly how many centimetres they have grown each year of their life. Explain that for the purposes of this activity, approximations will do just as well.
- You may wish the children to find out the measurements at home, and construct their height chart at school.
- If appropriate, children represent their data using ICT.

The Arts Roundup

- An important aspect of this investigation is how the method of obtaining music varies between different age groups. For this reason, children may need to undertake part of this investigation at home.
- It is also important that children justify their conclusions with evidence.
- If appropriate, children represent their data or write about their results using ICT.

AfL

- Tell me some of your statements about this graph / Venn diagram.
- How does your diagram convey the same data? Which is quicker and easier to read? Why?
- Tell me the results of your investigation.
- What conclusions can you draw?
- Why did you choose to record your results in this way?
- Talk me through what you did for each of these steps. Why did you make that decision?

Issue 35 — Statistics

Answers

Page 1

The Language of Maths
Diagrams and statements will vary. However, diagrams may look something similar to the following:

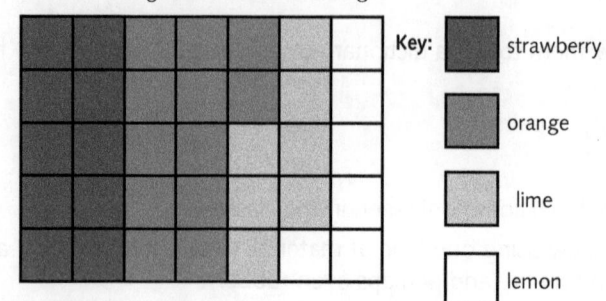

Key:
- strawberry
- orange
- lime
- lemon

Statements will vary.

At Home
Results of the investigation will vary.

Page 2

The Language of Maths
Statements will vary.

Let's Investigate
Results of the investigation will vary as will the form in which the data is represented.

Page 3

Let's Investigate
Results of the investigation will vary.

The Language of Maths
Lists will vary.

Focus on Science
Conclusions and evidence will vary.

Page 4

At Home
Height charts will vary.

The Arts Roundup
Answers and evidence will vary.

Inquisitive ant

conclusion
A decision or judgement formed by reasoning after considering the relevant facts or evidence.

Issue 36
Statistics

Prerequisites for learning

- Answer a question by collecting, organising and interpreting data
- Use tally charts, frequency tables, pictograms and bar charts to represent results and illustrate observations
- Use ICT to create a simple bar chart
- Use Venn diagrams or Carroll diagrams to sort data and objects using more than one criterion

Resources

pencil and paper
Resource sheet 2: My notes (optional)
Resource sheet 3: Pupil self assessment booklet (optional)
Resource sheet 8: 1 cm squared paper
Resource sheet 9: 2 cm squared paper
ruler
pack of playing cards
data handling software (optional)
computer with Internet access

Teaching support

Page 1

Around the World

- Ensure children are familiar with the five steps of the data handling cycle.
- An important aspect of this activity is the children's reflection, upon completion of the investigation, on each of the five steps of the cycle.
- You may need to discuss with some of the children the best way to collect, organise and represent the data.
- If appropriate, children represent their data using ICT.

At Home

- Once the children have completed the activity, ensure that there is an opportunity in class for pairs or groups of children to compare their lists and discuss the different criteria they used for organising their lists.
- Possible categories could be:
  - usage of appliance, e.g. cooking, cleaning, temperature control, entertaining, ...
  - location of appliance, e.g. kitchen, bathroom, sitting room, bedroom, ...
  - who uses the appliance most, e.g. mum, dad, my brothers and sisters, me, ...

Page 2

The Language of Maths

- Prior to setting the children to work independently on this activity, briefly discuss the two graphs with them, highlighting the labelled axes and their meanings. Ask the children to begin to tell the "story" of each graph.
- Ensure children realise that there is no one correct answer to this activity. What is important is that whatever interpretations and conclusions they make, they are able to justify them.

Focus on Science

- Ensure children can interpret the diagram correctly as some children may be unfamiliar with this type of Venn diagram.
- Ask the children to write statements describing their completed Venn diagram.

Page 3

Focus on Science

- Children will need to spend time on the Internet investigating different animals and their classification and habitat.

Issue 36 – Statistics

- Some children may need some assistance in deciding what criteria to use when constructing the second Carroll diagram.

Let's Investigate

- This activity is designed to introduce children to the concept of chance and likelihood, i.e. probability.
- Allow the children to interpret the activity for themselves and how to describe the different probabilities, i.e. using terms such as "in", "out of", "50–50"; or using fractions or even percentages.
- Once pairs have completed the activity, hold a group discussion where different pairs can share how they described each of the probabilities.

Page 4

The Language of Maths

- The names given to each of the 12 tables, charts and graphs may be slightly different to those used in some schools. It is suggested therefore that prior to children undertaking this activity that teachers look carefully at the issue to ensure that children will be able to interpret and, through a process of elimination, arrive at the correct labels even if they are not familiar with all the diagrams.
- Once children have completed the activity, arrange them into pairs to share and discuss their results.

AfL

- Tell me the results of your investigation.
- What conclusions can you draw?
- Why did you choose to record your results in this way?
- Tell me what you did for each of these steps. Why did you decide to do that?
- What criterion did you use to sort your list? What other criteria did you use?
- Tell me the "story" of these two graphs.
- Describe your Venn / Carroll diagram to me.
- How likely is it that you would turn over the Ace of Spades? What about a red card?
- What can you tell me about the likelihood of an event occurring?
- Tell me an event. How likely is it to happen? Tell me something that is certain to happen.
- Which graphs / charts were easy to name? Which were not so easy? What made you decide that it was called this?

Answers

Page 1

Around the World
Results of the investigation will vary.

At Home
Results of the investigation will vary.

Page 2

The Language of Maths
Statements will vary.

Focus on Science
Venn diagrams will vary.

Page 3

Focus on Science
Results of the investigation will vary.

Let's Investigate
Explanations will vary. However, they should be similar to the following:
- There is a one in four chance that the card would be the King of Hearts.
- There is a 50–50 chance that the card would be a black card.
- You are twice as likely to choose a red card over the Ace of Spades.

Page 4

The Language of Maths

Table	Line graph	Bar chart
Tally chart	Pictogram	Conversion graph
Block graph	Venn diagram	Frequency table
Pie chart	Scattergram	Carroll diagram

Inquisitive ant

pie chart
A pie chart shows groups of information as sectors of a circle. The larger the sector of circle, the larger the amount of data it represents.

Stretch and Challenge 3 Record of completion

Resource sheet 1

Class/Teacher: _____

Domain(s)	Topic	Stretch and Challenge Issue	Names									
Number: – Number and place value	Number	1										
	Number	2										
	Number	3										
	Number	4										
Number: – Addition and subtraction	Addition	5										
	Addition	6										
	Subtraction	7										
	Subtraction	8										
Number: – Multiplication and division	Multiplication	9										
	Multiplication	10										
	Division	11										
	Division	12										
Number: – Addition and subtraction – Multiplication and division	Mixed operations	13										
	Mixed operations	14										
	Mixed operations	15										
	Mixed operations	16										

© HarperCollins*Publishers* Ltd. 2016

Resource sheet 1

Domain(s)	Topic	Stretch and Challenge Issue	Names									
Number: – Fractions	Fractions	17										
	Fractions	18										
	Fractions	19										
	Fractions	20										
Measurement	Length	21										
	Mass	22										
	Capacity and volume	23										
	Time	24										
	Measurement	25										
	Measurement	26										
Geometry: – Properties of shapes	2-D shapes	27										
	3-D shapes	28										
	Symmetry	29										
	Position and direction	30										
	Movement and angle	31										
	Geometry	32										
	Geometry	33										
Statistics	Statistics	34										
	Statistics	35										
	Statistics	36										

© HarperCollins*Publishers* Ltd. 2016

Resource sheet 2

The Maths Herald

s&C Volume 3

Name:

Date:

Date of starting issue:

Date of finishing issue:

My notes

My notes

Inquisitive ant

Busy Ant Maths *Stretch and Challenge 3* © HarperCollins*Publishers* Ltd. 2016

Resource sheet 2

My notes

My notes

Busy Ant Maths *Stretch and Challenge 3* © HarperCollins*Publishers* Ltd. 2016

The Maths Herald

s&c Volume 3

Name:

Date:

Date of starting issue: Date of finishing issue:

What have you learned?

What did you use to help you?

Any other comments?

Your teacher's comments

Resource sheet 3

What did you enjoy the most?

What did you enjoy the least?

Did you find the work:
too easy?
just about right?
too hard?

What would you like to learn next about this topic and why?

Symmetrical circles

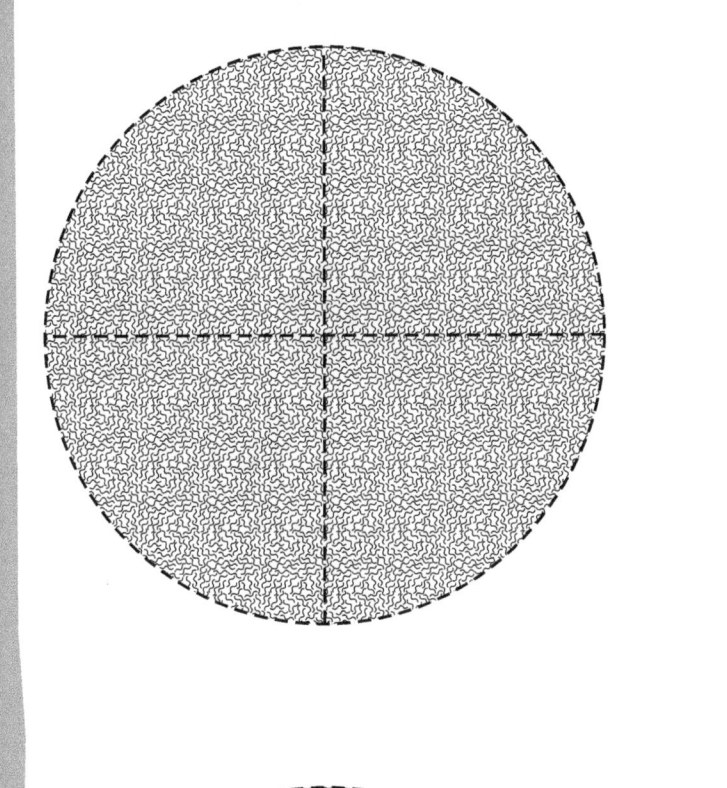

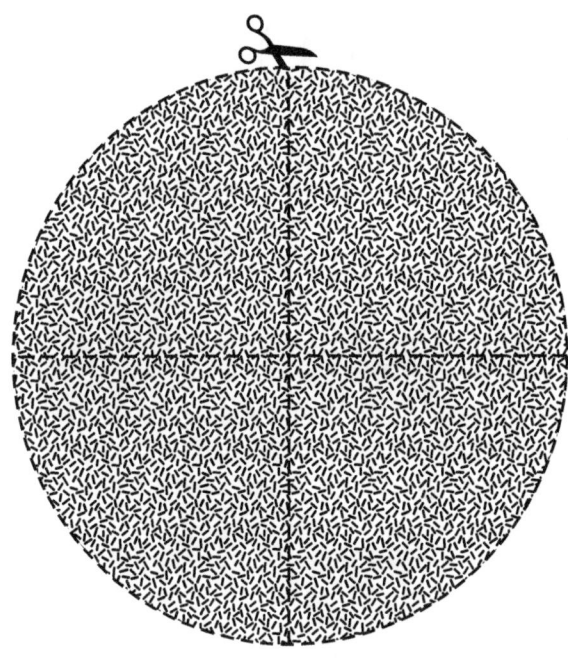

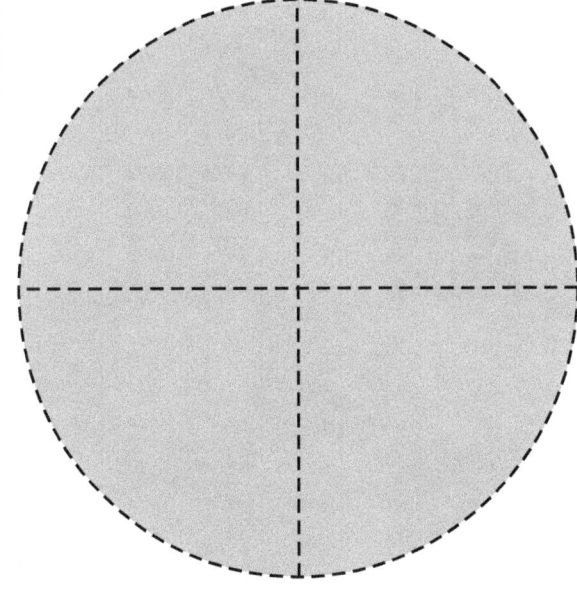

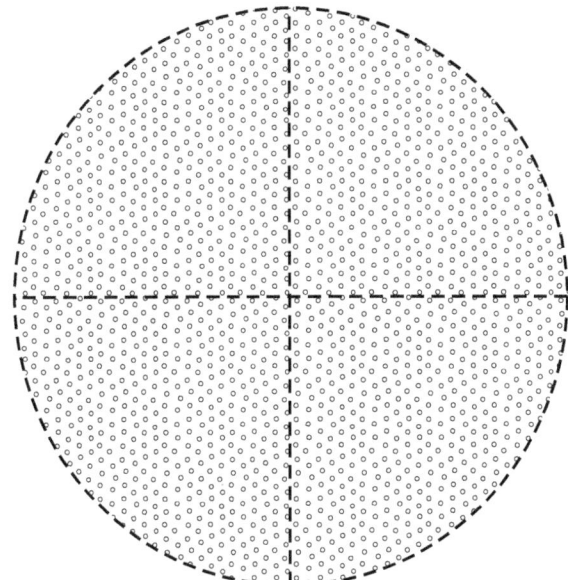

Resource sheet 4

Busy Ant Maths *Stretch and Challenge 3*

© HarperCollins*Publishers* Ltd. 2016

Draughts board

Resource sheet 5

① ②

③  ④

Busy Ant Maths *Stretch and Challenge 3* © HarperCollins*Publishers* Ltd. 2016

Half a shape

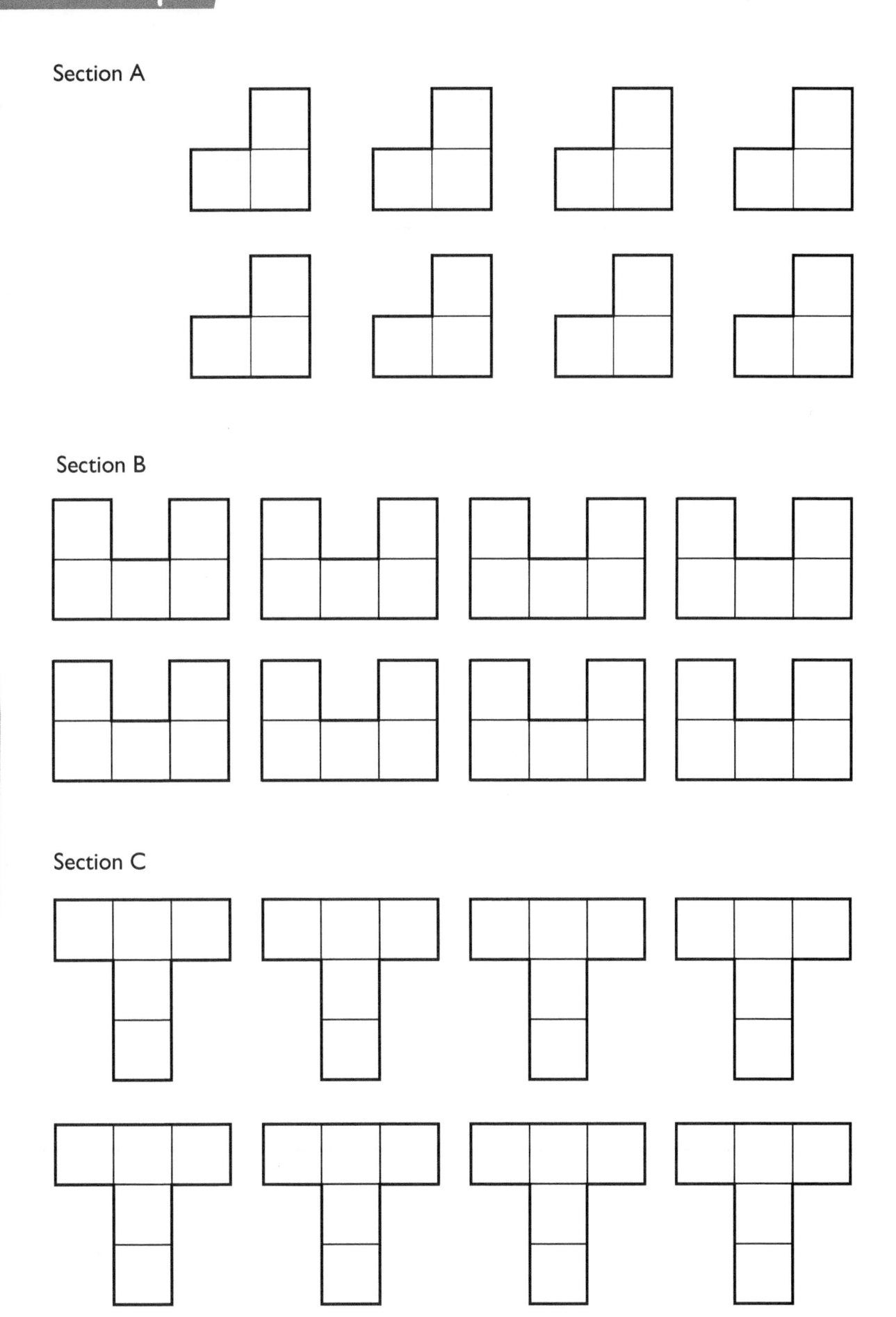

Resource sheet 6

Section A

Section B

Section C

Busy Ant Maths *Stretch and Challenge 3*

Tangram

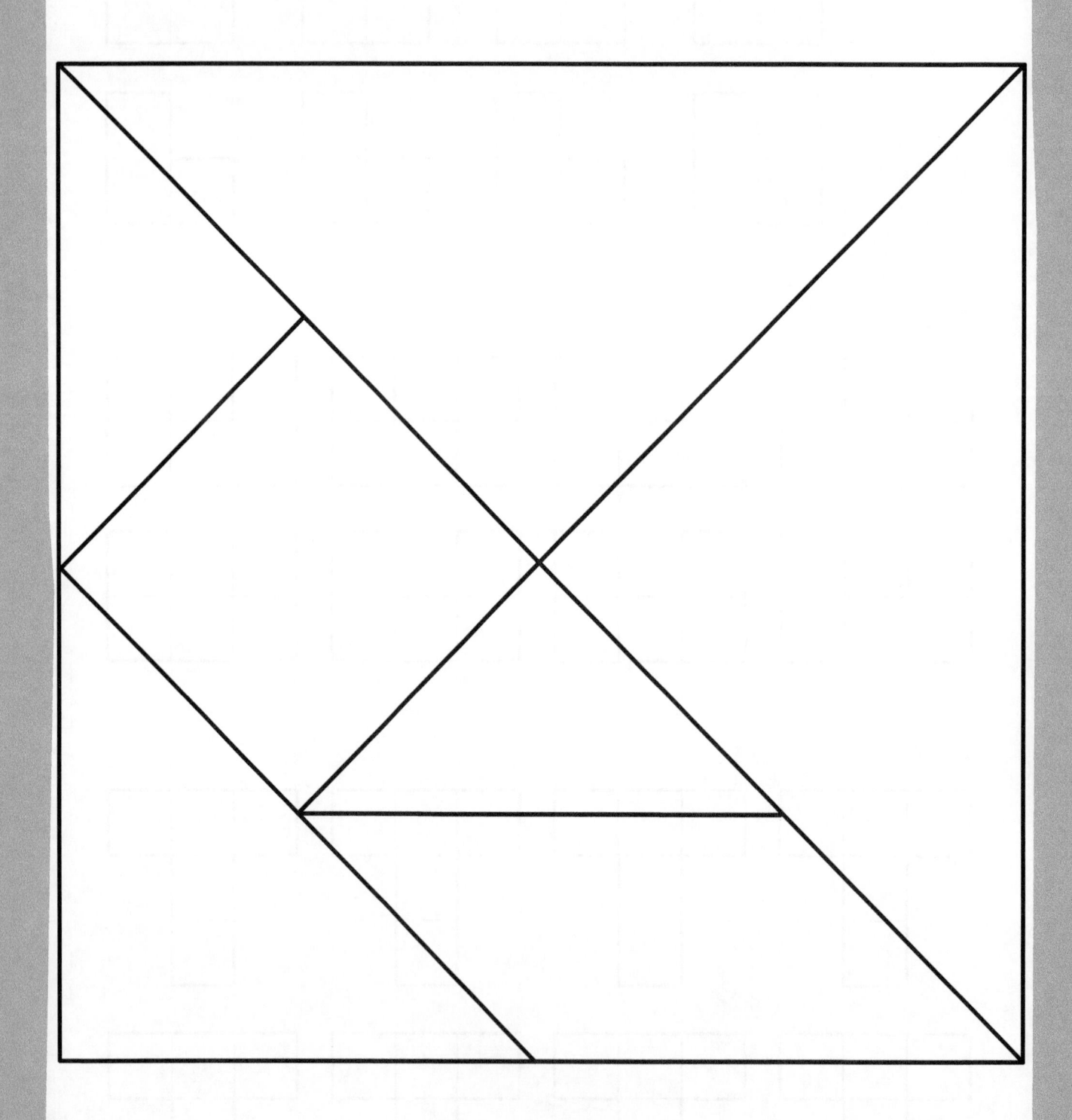

Resource sheet 7

Busy Ant Maths *Stretch and Challenge 3*

© HarperCollins*Publishers* Ltd. 2016

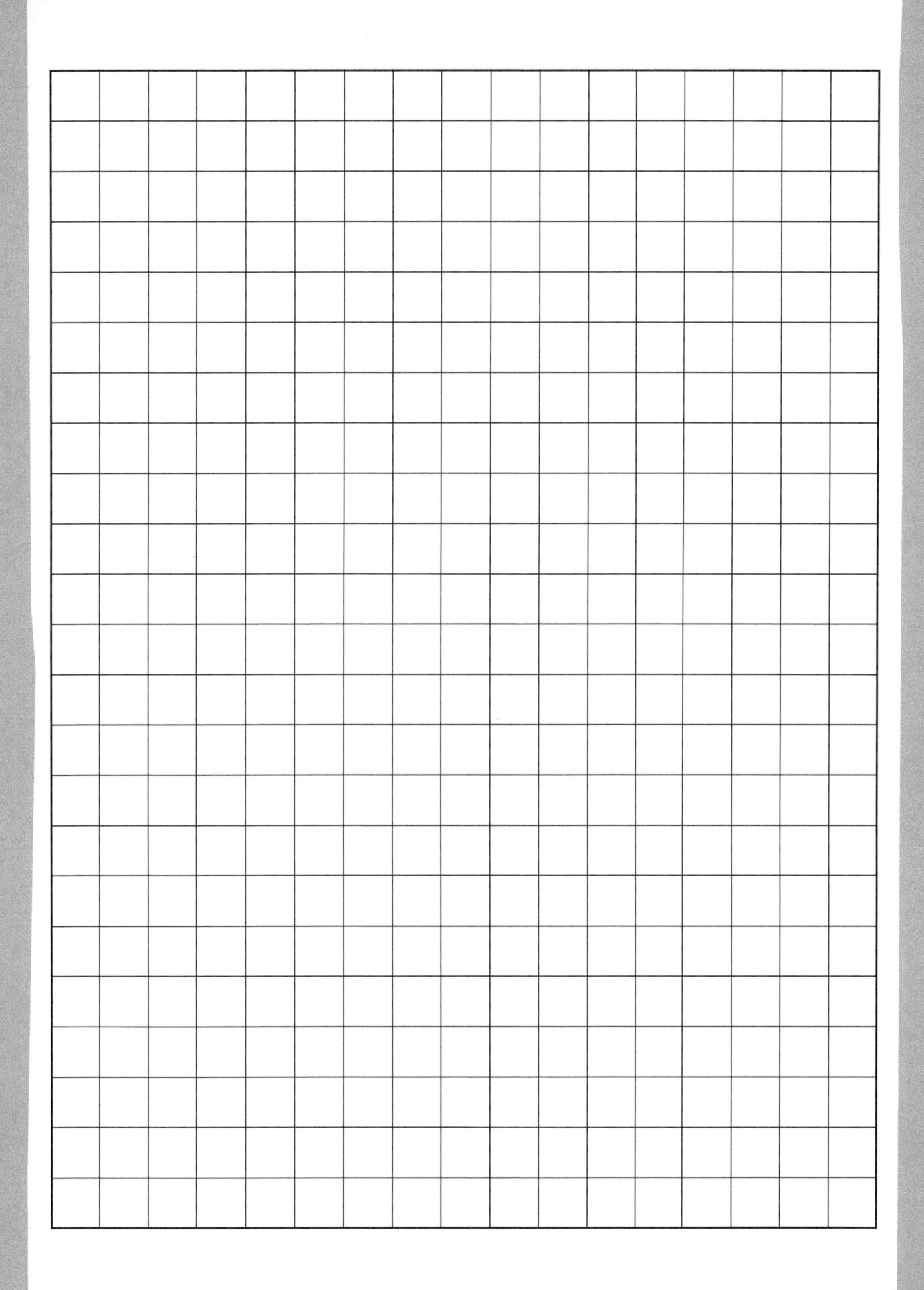

1 cm squared paper

Resource sheet 8

2 cm squared paper

0·5 cm squared paper

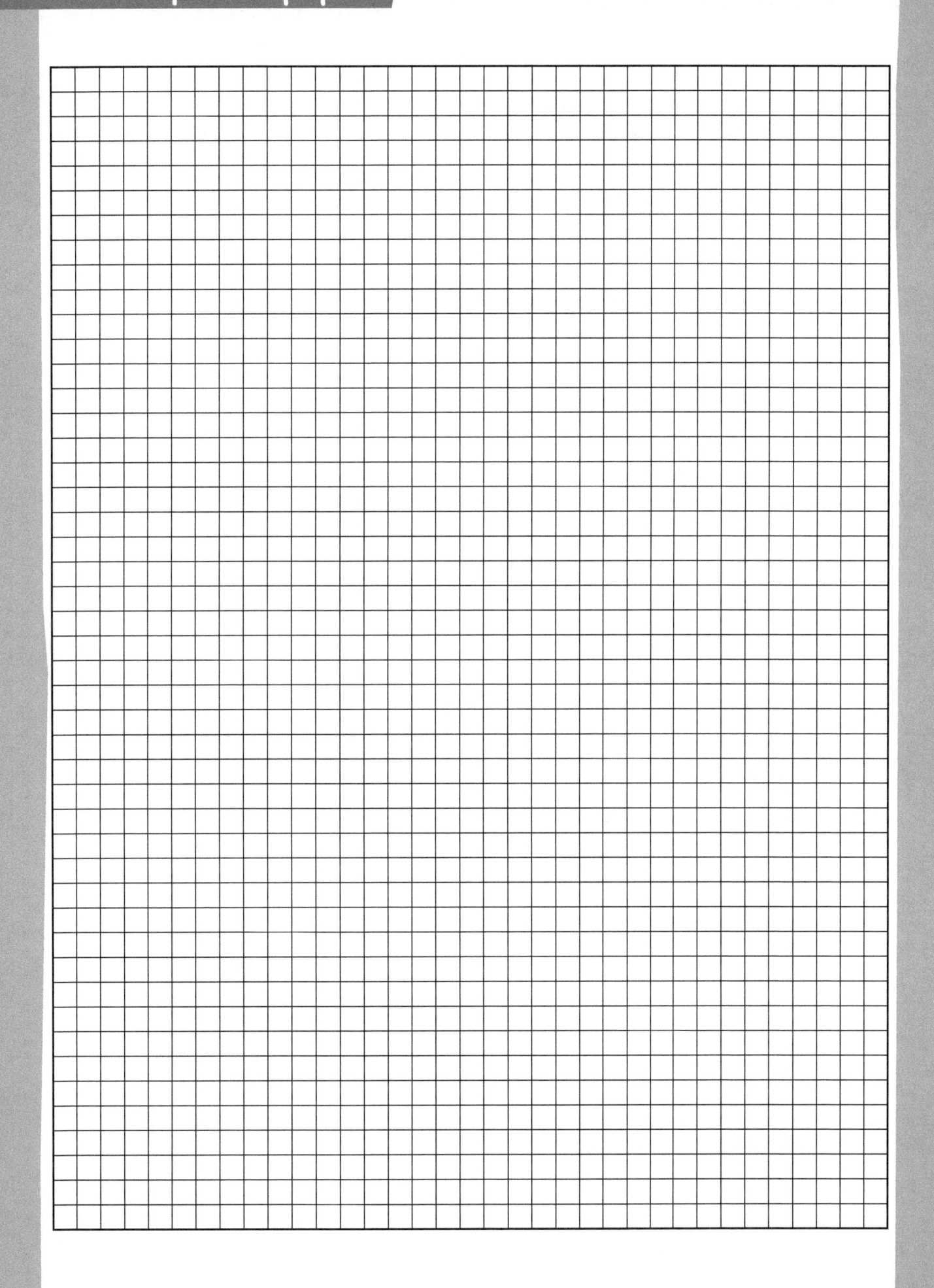

Resource sheet 10

Busy Ant Maths *Stretch and Challenge 3* © HarperCollins*Publishers* Ltd. 2016

Squared dot paper

Resource sheet 11

# Triangular dot paper

Resource sheet 12

Busy Ant Maths *Stretch and Challenge 3* © HarperCollins*Publishers* Ltd. 2016

Isometric paper

Resource sheet 13